Agriculture Issues and Policies Series

America's Family Farms

AGRICULTURE ISSUES AND POLICIES SERIES

Agriculture Issues & Policies, Volume I
Alexander Berk (Editor)
2001. ISBN 1-56072-947-3

Agricultural Conservation
Anthony G. Hargis (Editor)
2009. ISBN 978-1-60692-273-6

Agricultural Conservation
Anthony G. Hargis (Editor)
2009. ISBN 978-1-60876-813-4 (Online book)

Environmental Services and Agriculture
Karl T. Poston (Editor)
2009. ISBN: 978-1-60741-053-9

Weeds: Management, Economic Impacts and Biology
Rudolph V. Kingely (Editor)
2009. ISBN 978-1-60741-010-2

Soybean and Wheat Crops: Growth, Fertilization and Yield
Samuel Davies and George Evans
2009. ISBN: 978-1-60741-173-4

Effects of Liberalizing World Agricultural Trade
Henrik J. Ehrstrom (Editor)
2009. ISBN: 978-1-60741-198-7

Effects of Liberalizing World Agricultural Trade
Henrik J. Ehrstrom (Editor)
2009. ISBN: 978-1-60876-601-7 (Online book)

Hired Farmworkers: Profile and Labor Issues
Rea S. Berube (Editor)
2009. ISBN 978-1-60741-232-8

Agricultural Wastes
Geoffrey S. Ashworth and Pablo Azevedo (Editors)
2009. ISBN: 978-1-60741-305-9

Sugar Beet Crops: Growth, Fertilization & Yield
Claus T. Hertsburg
2009. ISBN: 978-1-60741-491-9

Economic Impacts of Foreign-Source Animal Disease
Jace R. Corder (Editor)
2009. ISBN: 978-1-60741-601-2

Economic Impacts of Foreign-Source Animal Disease
Jace R. Corder (Editor)
2009. ISBN: 978-1-60876-602-4 (Online book)

Essential Oils: Art, Agriculture, Science, Industry and Entrepreneurship (A Focus on the Asia-Pacific region)
Murray Hunter
2009. ISBN: 978-1-60741-865-8

Corn Crop Production: Growth, Fertilization and Yield
Arn T. Danforth
2009. ISBN: 978-1-60741-955-6

Corn Crop Production: Growth, Fertilization and Yield
Arn T. Danforth
2009. ISBN: 978-1-60876-860-8 (Online book)

Phosphate Solubilizing Microbes for Crop Improvement
Mohammad Saghir Khan and Almas Zaidi (Editors)
2009. ISBN: 978-1-60876-112-8

Organic Food - Economics and Issues
Earl D. Straub (Editor)
2009. ISBN: 978-1-60741-130-7

Global Beef Trade
Alessandro Ferrara (Editor)
2009. ISBN: 978-1-60741-121-5

Ecophysiology of Tropical Tree Crops
Fabio DeMatta (Editor)
2010. ISBN 978-1-60876-392-4

Assessing Disease Potential in U.S. Aquaculture Industry
Emilija Kalnins (Editor)
2010. ISBN: 978-1-60741-543-5

Inspection and Protection of U.S. Meat and Poultry
Finn J. Amundson (Editor)
2010. ISBN: 978-1-60741-120-8

Soil Phenols
A. Muscolo and M. Sidari (Editors)
2010. ISBN: 978-1-60876-264-4

No-Till Farming: Effects on Soil, Pros and Cons and Potential
Earl T. Nardali (Editor)
2010. ISBN: 978-1-60741-402-5

Drivers and Restraints for Economically Efficient Farm Production
Helena Hansson Karin Larsén Bo Öhlmér
2010. ISBN: 978-1-60876-171-5

Agriculture and Food
Claus Schäfer (Author)
2010. ISBN: 978-1-60692-038-1

Agriculture Research and Technology
Kristian Bundgaard and Luke Isaksen (Editors)
2010 ISBN: 978-1-60741-850-4

Agricultural Economics: New Research
Tomas H. Lee (Editor)
2010. ISBN: 978-1-61668-077-0

U.S. Biobased Products Market Potential and Projections through 2025
Meredith A. Williamson (Author)
2010 ISBN: 978-1-60741-033-1

The Sugar Industry and Cotton Crops
Peter T. Jenkins (Editor)
2010. ISBN: 978-1-61668-320-7

Tomatoes: Agricultural Procedures, Pathogen Interactions and Health Effects
Eric D. Aubé and Frederick H. Poole (Editors)
2010. ISBN: 978-1-60876-869-1

Organic Farming and Peanut Crops
Darren C. Grossman and Terrance L. Barrios (Editors)
2010. ISBN: 978-1-60876-187-6

Farm Bill of 2008: Major Provisions and Legislative Action
Mary T. Conner (Editor)
2010. ISBN: 978-1-60741-750-7

America's Family Farms
Efren J. Tamayo (Editor)
2010. ISBN: 978-1-60741-751-4

Pesticide Resistance, Population Dynamics and Invasive Species Management
Gregory J. McKee, Colln A. Carter, James A. Chalfant, Rachael E. Goodhue, and Frank G. Zalom (Authors)
2010. ISBN: 978-1-60741-758-3

Rethinking Structural Reform in Turkish Agriculture: Beyond the World Bank's Strategy
Baris Karapinar, Fikret Adaman and Gokhan Ozertan (Editors)
2010. ISBN: 978-1-60876-718-2

Manure Use for Fertilizer and Energy
Connor D. Macias (Editor)
2010. ISBN: 978-1-60876-847-9

Price Dynamics behind Consumer Food Purchases
Morgan D. Fitzpatrick (Editor)
2010. ISBN: 978-1-60876-892-9

Governance of Agrarian Sustainability
Hrabrin Bachev (Author)
2010. ISBN: 978-1-60876-888-2

Transformation of U.S. Animal Agriculture
Justin M. Daigle (Editor)
2010. ISBN: 978-1-60876-938-4

The Peanut Plant and Light: Spermidines from Peanut Flowers
Victor S. Sobolev, James B. Gloer and Arlene A. Sy (Authors)
2010. ISBN: 978-1-61668-028-2

Agriculture and Environmental Security in Southern Ontario's Watersheds
Glen Filson, Bamidele Adekunle, and Katia Marzall (Authors)
2010. ISBN: 978-1-61668-156-2

Sustainable Agriculture: Technology, Planning and Management
Augusto Salazar and Ismael Rios (Editors)
2010. ISBN: 978-1-60876-269-9

America's Family Farms
Efren J. Tamayo (Editor)
2010. ISBN: 978-1-60741-751-4

AGRICULTURE ISSUES AND POLICIES SERIES

AMERICA'S FAMILY FARMS

EFREN J. TAMAYO
EDITOR

Nova Science Publishers, Inc.
New York

For permission to use material from this book please contact us:
Telephone 631-231-7269; Fax 631-231-8175
Web Site: http://www.novapublishers.com

LIBRARY OF CONGRESS CATALOGING-IN-PUBLICATION DATA

America's family farms / editor, Efren J. Tamayo.
p. cm.
Includes index.
ISBN 978-1-60741-751-4 (hardcover)
1. Family farms--United States. I. Tamayo, Efren J.
HD1476.U6A44 2009
338.10973--dc22

2009032002

Published by Nova Science Publishers, Inc. † New York

CONTENTS

PREFACE

This book explores America's family farms today. Farmers face many challenges like making daily decisions on when to plow, till, plant, fertilize, and harvest, or management decisions involving their sizable assets, numerous employees, a multitude of rules and regulations and a market place that changes every day. In addition to these challenges, competition from foreign markets, environmental concerns, low prices and many global factors also make the future uncertain. We are at a critical stage in planning for the future of agriculture; recruiting and training the next generation of farmers and ensuring farms will continue to be viable, healthy operations. Every family owned business has to deal with transitions, but it seems to be reaching crisis proportions in agriculture.

Chapters 1 through 4 feature testimony before the U. S. Senate Special Committee on Aging.

Chapter 5 - American farms encompass a wide range of sizes, ownership structures, and business types. A farm classification developed by USDA's Economic Research Service (ERS) categorizes farms into more homogeneous groupings for reporting and evaluation purposes. The typology is based largely on farm sales, organization, and the farm operator's primary occupation.

Chapter 6 - Farm subsidy programs in the 1930s were largely prompted by concern for the chronically low, and highly variable, incomes of U.S. farm households. Seventy years later, commodity-based support programs are still prominent, even though the income and wealth of the average farm household now exceed those of the average nonfarm household—wealth by a large margin.

Farm households continue to face variability in income due to weather and natural disasters. Household income is most variable for the small segment that operates commercial farms (above $250,000 in annual sales). Relative to small farms, these farms achieve greater economies of scale, generate higher profit margins, and their households realize a larger share of their income from farming. However, the substantial net worth of these households acts as a cushion against uncertain farm income, much as off-farm income does for households operating smaller farms.

In a variable-income/high-wealth sector such as farming, economic well-being measures based on both income and wealth can provide a better signal of household capacity to support a consistent living standard than income measures alone. In 2003, 5 percent of farm households had both income and wealth below the respective U.S. household medians, and those households, on average, spent more on basic consumption than they earned in income.

Households with low income and low wealth are less likely to receive farm payments, excluding conservation programs; by contrast, only 3 percent of households receiving payments had income and wealth below the U.S. household medians for each.

Chapter 7 - Commodity program payments shifted sharply to higher income households between 1989 and 2003. While Congress made important changes to program design in the 1996 and 2002 farm bills, this shift was not caused by change in design. Instead, changes in payment flows resulted from structural changes in farming that are driving production to very large family farms. We expect those structural changes to continue, because larger farms appear to be more profitable and because many more operators of smaller farms are nearing retirement age. In consequence, in the absence of any fundamental changes to commodity policy, commodity payments will continue to shift to higher income households.

Chapter 8 - The economic well-being of most U.S. farm households depends on income from both onfarm and off-farm activities. Consequently, for many farm households, economic decisions (including technology adoption and other production decisions) are likely to be shaped by the allocation of managerial time among such activities. While time allocation decisions are usually not measured directly, we observe the outcomes of such decisions, such as onfarm and off-farm income. This report finds that a farm operator's off-farm employment and off-farm income vary inversely with the size of the farm. Operators of smaller farm operations improve their economic performance by compensating for the scale disadvantages of their farm business with more off-farm involvement. Off-farm work reduces farm-level technical efficiency, but increases household-level technical efficiency. And adoption of agricultural innovations that save managerial time is associated with higher off-farm income.

Chapter 9 - U.S. farms are diverse, ranging from small retirement and residential farms to enterprises with annual sales in the millions. Nevertheless, most U.S. farms—98 percent in 2004—are family farms. Even the largest farms tend to be family farms. Large-scale family farms and nonfamily farms account for 10 percent of U.S farms, but 75 percent of the value of production. In contrast, small family farms make up most of the U.S. farm count, produce a modest share of farm output, and receive substantial off-farm income. Many farm households have a large net worth, reflecting the land-intensive nature of farming.

Chapter 10 - Despite high production costs, 2008 represented another year of record profitability for the U.S. farm economy as a whole. According to USDA's Economic Research Service (ERS), national net farm income—a key indicator of U.S. farm well-being—rose to a record $86.9 billion in 2008, marginally above the previous year's record ($86.8 billion). The growth in cash receipts to a record $323.4 billion for crop and livestock sales (up $38.6 billion or 14% from 2007) was nearly offset by a surge in production costs (up $38.2 billion or 15%) to a record $292.5 billion.

Government farm payments are projected up slightly in 2008 at $12.5 billion. An increase in ad hoc disaster payments more than offset lower commodity program payments, which declined when high crop prices rose above the price triggers for marketing loan benefits and countercyclical payments in 2008.

Within the farm balance sheet, total farm asset value of $2,359 billion and total farm debt of $212 billion are both projected at record levels in 2008. The debt-to-asset ratio of 9.0% is down sharply from last year's value of 9.6% and represents the lowest level since 1960, suggesting a strong financial position for the agricultural sector as a whole.

However, less than ideal market conditions heading into 2009 suggest dim prospects for the longer-term farm income outlook, albeit surrounded by considerable uncertainty. On the one hand, the global financial crisis, economic recession, rising unemployment, limited credit availability, and plummeting asset values that persist in early 2009 have contributed to substantial "demand destruction" (i.e., a severe weakening of consumer demand), which bodes poorly for farm commodity price prospects. On the other hand, weak energy markets and declining input prices could provide some spark to both producer investment and consumer demand for agricultural sector products, perhaps by the middle to latter half of the year. USDA will release its first U.S. farm income forecasts for 2009 on February 12, 2009.

In: America's Family Farms
Editor: Efren J. Tamayo

ISBN: 978-1-60741-751-4

Chapter 1

AGING FARMERS HEARING- BUSHUE TESTIMONY*

Barry Bushue

Chairman Kohl, Ranking Member Smith and members of the committee, I want to thank you for the opportunity to testify before you today. My name is Barry Bushue and I am currently serving as President of the Oregon Farm Bureau Federation (OFBF). It is a privilege to represent all members of the Oregon Farm Bureau and the agriculture community in the great State of Oregon.

OFBF is the largest agriculture organization in the state, representing all aspects of the industry. Consequently our interests are wide and diverse. However, among the many interests of farmers and ranchers in Oregon and across the US, one key element ties them together. That is the desire to keep agriculture economically viable, passing down family operations to future generations and to continue to provide American's with the lowest cost of food and fiber in the world.

More often than not, the general public regards the farm bill as the only legislation affecting the industry. What most don't understand is that nearly every aspect of public policy has a significant impact on agriculture. These issues, outside the scope of the farm bill, and the effect on future or beginning farmers are what I would like to focus on today.

As an ambassador for Oregon agriculture I am deeply concerned about the future of the industry. Farming is not easy. There are no guarantees, no paychecks every two weeks, little stability and it is extremely expensive to start out. These statements probably beg the question, why in the world would anyone want to be part of the industry? The answer is simple. It is a shear love of the land. Environmental conservation is of the utmost importance to farmers and ranchers. I am thankful everyday that I am able to enjoy Oregon's beautiful surroundings. It is too often that farmers and ranchers are labeled as anti-environmentalists, but the exact opposite is the case. Farmers and ranchers are directly involved in the protection

* This is an edited, reformatted and augmented edition of testimony before the Senate Committee, Special Committee on Aging, submitted by Barry Bushue on June 21, 2007.

and utilization of the environment surrounding us. Preservation of land and conservation of other natural resources is necessary if we are to live and professionally thrive.

However, increasing and unnecessary regulations on the industry continue to force established farmers and ranchers out of business. How is a young person supposed to cope? For example, the Agricultural Protection and Prosperity Act of 2007, currently circulating through Congress regarding the Comprehensive Environmental Recovery, Compensation & Liability Act is critical to future producers. I do not believe it was the intention of Congress for CERCLA to apply to manure. The law clearly exempts the application of chemical fertilizers containing the same constituents as manure, which occur naturally in the environment. Furthermore, animal agriculture operations are already regulated under the Clean Water Act, the Clean Air Act and various state laws to protect the environment

Regulations in areas such as animal agriculture are also a critical disincentive for new entrants and encourage aging operators to exit earlier. For example, over the last decade, Illinois lost half of its livestock operators and the burden and costs associated with environmental regulation were likely a factor in those decisions. In many cases, this loss took the form of a 55 year old operator exiting now rather remodeling facilities and changing operating procedures radically. Most of what would have been the natural replacement of exiting operators was also cut off by the complicated regulations governing new entrants.

The agriculture industry continues to educate Congress on animal rights groups, their strategies and their goal of ending animal production for food. Not only would this have a devastating economic impact, but also the public campaigns on behalf of these groups provide with public with misinformation and half truths. The mission of our industry is to educate Congress and the public on animal rights issues, as well as provide tools to assist in state and local animal rights legislative and regulatory challenges.

This is not a large versus small farm issue or better put, a corporate versus family farm issue. The statutes current reporting requirements and liability thresholds for non-ag release and/or emissions of regulated substances are extremely low. Virtually any agricultural operation producing, storing and/or using animal manure could be held liable.

Oregon is a specialty crop state. We are famous worldwide for our fruits and vegetables as well as our nursery products. It is a very competitive and fragile industry impacted by spikes in plantings, production and weather. Most importantly, the availability of labor is critical to a successful operation. Again, the unpredictability and uncertainly is certainly not an enticement for young people.

America's farmers and ranchers face a "Catch 22" when verifying the status of their workforce. It is illegal to knowingly to hire someone who is not authorized to work, but the employer is limited in what he or she may ask to determine who is authorized. If the employer requests more or different documents when the original documents appear reasonable on their face, the employer could be subject to Justice Department investigation or lawsuits for discriminating in employment. If, on the other hand, the employer accepts those documents but later is notified by the Social Security Administration that information contained in the documents does not match agency records, then the employer may not be safe from prosecution for knowingly hiring an illegal worker.

I commend the efforts of the Senate regarding bipartisan comprehensive immigration reform. Your hard work would provide immediate relief from labor shortages. It is critical that comprehensive immigration reform legislation get signed into law this year. If, however, President Bush signs into law a bill that does not fully address agriculture's labor needs, economists estimate that up to $9 billion nationwide in annual production is at risk of being lost to foreign competition. Oregon agriculture is estimated to lose nearly $5 million. A successful comprehensive immigration reform must include a reliable verification system, adequate transition provisions and a viable guest worker program.

The estate tax is a tremendous burden on farmers and ranchers. Individuals, family partnerships or family corporations own 99 percent of the two million farms that dot America's rural landscape. Farms face a common problem of being land rich, but cash poor. The burden of the estate tax, which can be as a high as 45 percent, too often forces young farmers and ranchers to sell land, buildings or equipment needed to operate their businesses. When farms and ranches disappear, the rural communities and businesses they support also suffer.

The average estate tax payment in 1999 to 2000 was the equivalent of one-and-a-half to two years of net farm income. Roughly twice the number of farm estates paid federal death taxes compared to other estates in the late 1990s. Moreover, the average farm death tax is also larger than the tax paid by most other estates. Heirs should have the choice of valuing land at either fair market value or current use value without limitation, and there should be no estate tax on land that remains in agricultural production.

Congress voted to end death taxes in 2001. The law provided immediate relief through rate reduction and an expanded exemption, with complete repeal occurring in 2010. Unfortunately, the bill's provisions expire in 2011, requiring Congress to pass additional legislation to make death tax elimination permanent. I urge the Senate to take up companion legislation to HR 2380, permanently repealing the inheritance tax.

Finally, I would be remiss if I didn't mention the reauthorization of the Secure Rural Schools Act is essential. This funding is vital for the education of children in Oregon and across the country. More than 100 years ago, Congress recognized that these counties and communities with federally owned forestland within their boundaries would not be able to provide basic services because of the reduction in local property tax base. Congress also recognized that national forests exist for the benefit of an entire nation, and therefore, the entire nation has a role in maintaining the health of our forests and their surrounding communities.

This is an exciting time for US agriculture. So much of what happens in Congress has a direct impact on the industry. As farmers and ranchers continue to utilize cutting edge technology, yields will increase while maintaining our international reputation of providing American's with the lowest cost of food in the world. . It is with sincere gratitude that I thank you for not only the opportunity to share with you the many challenges facing young farmers and ranchers, but also for the great work you do day in and day out.

In: America's Family Farms
Editor: Efren J. Tamayo
ISBN: 978-1-60741-751-4

Chapter 2

AGING FARMERS HEARING- CONNOR TESTIMONY*

Charles Conner

Mr. Chairman and Committee members, thank you very much for the opportunity to discuss the issues related to older farmers and ranchers in American agriculture. Today, I will provide a brief overview of the demographic and financial profile of U.S. farmers and identify some of the challenges that an aging farm workforce presents. I will then discuss the programs the Department of Agriculture operates and the Administration's 2007 Farm Bill proposals that address these challenges.

GENERALLY PROSPEROUS TIMES FOR U.S. AGRICULTURE

Before addressing specific issues related to age, there are several developments in agriculture that define the context in which age issues will play out over the next several years. A key factor is the strengthening of the farm economy. After a slowdown in the late 1990s and early 2000s following the sharp devaluation of Asian currencies and the global slowdown in economic growth, U.S. agriculture has enjoyed an unprecedented period of economic prosperity. Aggregate net cash farm income reached a record high $81.5 billion in 2004, followed by a near record $81.2 billion in 2005. Income declined to an estimated $67 billion in 2006, as government payments declined and production expenses, especially energy-related costs, increased. The significance of these numbers can be appreciated by comparing them to $64.6 billion which is the 10-year average of net cash farm income from 1997 to 2006. Another above average income year is forecast for 2007.

Key factors contributing to the robust farm economy include strong global economic growth and the rising use of agricultural products in biofuels. Strong foreign income growth is expected to help boost U.S. agricultural exports to a record high $77.5 billion in fiscal year (FY) 2007, up 13 percent from FY 2006. The $8.8 billion annual increase would be the second highest ever. Along with strong export demand, the demand for biofuels is expected to

* This is an edited, reformatted and augmented edition of statements made before the Senate Committee, Special Committee on Aging, by Charles Conner, Deputy Secretary of Agriculture on June 21, 2007.

drive corn prices to record-high levels for the 2007 crop. Ethanol production during the 2007/08 corn marketing year is expected to increase 58 percent above the 2006/07 level. Of total 2007 corn production, 27 percent is expected to be used in ethanol production, up from 20 percent of the 2006 production. Continued growth in biofuel production is expected to keep crop markets strong.

As income has improved, the wealth of U.S. farmers has also grown. At the start of 2007, U.S. farm net worth (assets minus debts) totaled $1.69 trillion, up from $1.18 trillion at the start of 2004. Another strong increase is expected in 2007. The major contributors to this growth have been modest increases in farm debt and substantial increases in farm real estate values. From 2004 to 2006 the average value of farm real estate has increased from $1,360 per acre to $1,900 per acre, a 40-percent increase in only 3 years.

AGING FARMERS

Now let's examine the issue of the age of U.S. farmers and their related economic well being. The average age of the U.S. population has been steadily rising for a long time. It increased especially rapidly over the past two decades as a result of the aging of the baby-boom generation. The average age of primary farm operators reflects this general population trend and also continues to rise, increasing by nearly 5 years from 50.5 in 1982 to 55.3 in 2002, the latest Census of Agriculture data. The Current Population Survey ranks all occupations by age for 2006, with farmers and ranchers averaging 55.9 years, the 6th highest of over 400 different occupations identified.

For nearly one-fourth of all farms with over $10,000 in annual sales, the primary operator was over 65 years of age and older in 2002 (Figure 1). In contrast, the Bureau of Labor Statistics estimates that only 8 percent of self-employed persons in nonagricultural industries were in that age category. The rise in the average age of primary farm operators reflects both a decline in the number of young farmers and a rise in the number of older farmers. From 1982 to 2002, the number of primary operators under 35 years of age declined while the number of farmers 65 years of age and older increased. Farms with over $10,000 in annual sales and whose primary operator was under 35 years of age accounted for less than 7 percent of all farms in 2002 as compared to 19 percent in 1982. Meanwhile, the share of farms with over $10,000 in sales whose primary operator was over 65 increased from 14 to 25 percent.

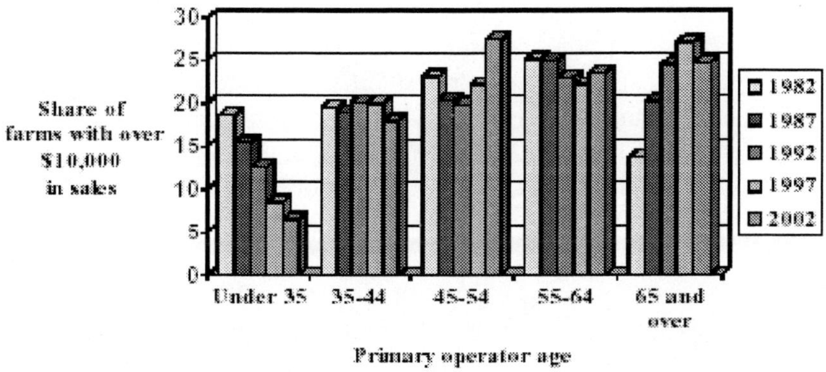

Figure 1. Farm Primary Operators by Age

Not surprisingly, older farmers are, in general, better off financially than younger farmers because they have had the time to accumulate assets and repay debts. In 2005, farm operators 65 years of age and older had average per farm assets of $775,687 compared to only $23,466 in farm liabilities; or a debt-to-asset ratio of 3 percent. Farm real estate averaged $660,681 on those farms, accounting for 85 percent of total assets. Alternatively, farm operators under 35 years of age had average assets of $480,261 per farm and $86,757 in farm liabilities; or a debt-to-asset ratio of 18 percent. For the farm sector as a whole, average per farm assets were $784,392 and farm liabilities were $62,127, for a debt-to-asset ratio of about 8 percent.

While farm operators 65 years of age and older may have a better balance sheet, they earn less income from farming compared with farmers less than 65 years of age. In 2005, the average net farm income for farms operated by a person 65 years of age and older was $18,179, compared with $28,361 for farms operated by someone under 35 years of age. For the farm sector as a whole, average net farm income per farm was $27,137 in 2005. Similar to other segments of society, these data show an older population of farmers who control a considerable amount of assets but earn less income from farming compared with younger farmers who have more debt but earn on average more income from farming.

Farm Succession Issues

The continuing decline in the number of farmers 34 years of age and younger has raised concerns that an insufficient pool of new entrants will be available to replace a large and growing pool of retiring farmers. There is no evidence, however, that a shortage of farm operators and farm workers has caused or will cause reduced production and higher prices of U.S. farm commodities. A USDA Economic Research Service study released in 2006 on farm exits (farms going out of business) indicated that while the total number of U.S. farms has declined very little in the 5 years between each Census of Agriculture, about 9 to 10 percent of farms go out of business each year. This suggests the apparent stability in the number of U.S. farms masks a rather dramatic turnover. Turnover rates are highest for smaller farms, but even for farms with sales of $250,000 or more, 6 to 7 percent go out of business each year. Thus, U.S. agriculture has been facing a substantial turnover for many years, and as these farms have been sold, new farm businesses have taken their place or existing farms have gradually increased in size.

There is also evidence that there are a substantial number of young farmers present on farms, though not as the primary operator. Data indicate that many commercial-sized farms with older operators also have younger operators involved in the farming operation. These secondary operators in many cases represent future primary operators. A high percentage of these farms are operated as partnerships and as family corporations. The 2002 Census of Agriculture indicated there were 194,000 farms, 9 percent of all farms, which have multiple operators with operators across different generations. In addition, the Current Population Survey identifies "farm, ranch, and other agricultural managers," who are paid to supervise and manage farm operations, as having an average age of 48.3 in 2006, 7.6 years less than the average age of farm operators.

Finally, the rising average age of primary operators may also reflect technology change that has enabled older farmers to more readily meet the physical demands of farm labor. Technology has reduced the time needed for field operations and has supplanted manual labor with mechanization, allowing farm size to increase over time and operators to spend more time on other farm management tasks. Improving health, reflected in longer lives, also enables the workforce in general to work to a much older age.

While there appears to be no impending slowdown in farm production resulting from the advancing age of primary operators, there are concerns about how upcoming intergenerational transfers may affect the future structure of agriculture. Older farmers hold a large share of farm assets. Primary operators 65 years of age and older owned over one-fourth of farm assets in 2004. Collectively, landowners 65 years of age and older owned over one-third of the total acres of land in farms. Ultimately, this pool of assets will either be sold or passed on to heirs. To the extent these assets are sold or leased to existing operators, this transfer raises concerns about consolidation and its effects on the structure of agriculture, local economies, and rural landscapes. Middle-sized family farms with annual sales of between $50,000 and $500,000 fell from 526,000 farms in 1982 to 381,000 in 2002. Meanwhile the number of large farms with over $500,000 in sales increased form 28,000 to 71,000 farms. This decline in middle-sized farms is one factor behind policies to encourage entry into agriculture or assist in the intergenerational transfer of farm assets. The concern about structure must be balanced against the economic effects of technology advancement and economies of scale, which enable agricultural demand to be met by larger, more efficient, and fewer farms.

Another factor likely to affect the transfer of farm assets is the cost of farm production assets. In 2002, the value of land and buildings averaged $710,000 for U.S. farms that were principally engaged in agriculture. For farms where the principal operator was under 35 years of age, the value of land and buildings averaged $595,000. Farmland values have been steadily increasing for many years, but jumped 15 percent in 2005 after a 21 percent increase in 2004, adding to the cost of entering farming. Thus, these figures indicate that access to capital for the purchase of land, buildings, and equipment may be a significant hurdle for many young farmers. However, renting land is a key option for young farmers, as nearly half of land in production is now rented. While the average value of farm real estate increased 57 percent between 2002 and 2006, the average rental rate on crop land only rose 10 percent.

USDA PROGRAMS AND ACTIVITIES

Uncertainty over farm succession and future farm structure have motivated programs to encourage individuals entering the workforce to take up careers as farmers and ranchers and help them succeed. The Agricultural Credit Improvement Act of 1992 was the first law that required USDA to provide special assistance to beginning farmers and ranchers. The law required the Secretary to establish: (1) beginning farmer loan programs; (2) Federal-State Beginning Farmer Partnerships for the purpose of providing joint financing to beginning farmers and ranchers; and (3) an Advisory Committee on Beginning Farmers and Ranchers. The Act also required that loan funds be targeted to beginning farmers and ranchers.

USDA implemented beginning farmer and rancher loan programs in FY 1994. From then through May 2007, the Farm Service Agency (FSA) has made more than 106,000 loans to beginning farmers and ranchers, totaling $9.5 billion. In FY 2006, 34 percent of all FSA direct and guaranteed loans have gone to beginning farmers and ranchers. FSA has also created Federal-State Partnerships by signing Memorandums of Understanding with 20 State beginning farmer programs, agreeing to provide joint financing to beginning farmers and ranchers. The Farm Security and Rural Investment Act of 2002 (2002 Farm Bill) authorized the Secretary to establish a Beginning Farmer and Rancher Land Contract Guarantee Pilot Program in not fewer than five states during FYs 2003-2007. The program, implemented in September 2003, provides FSA guarantees on loans made by private sellers of a farm or ranch on a contract land sales basis to qualified beginning farmers and ranchers. While available in nine states, only two guaranteed loans have been made since implementation.

While not a USDA agency, the Farm Credit System (FCS) also makes loans to young farmers and ranchers (35 years old and younger) and beginning farmers and ranchers (operating for not more than 10 years). For example, in 2006, FCS institutions made 58,000 loans with a total loan dollar value of $9.3 billion to beginning farmers and ranchers. This lending represented 21 percent of the total number of new loans and 18 percent of total dollar volume of new loans made by FCS institutions in 2006. As of December 31, 2006, loans to beginning farmers and ranchers accounted for almost 24 percent of the total number of loans outstanding and 19 percent or $25.4 billion of the dollar value of loans outstanding in the FCS.

USDA established the Advisory Committee on Beginning Farmers and Ranchers (Advisory Committee) in 1998. The Advisory Committee provides advice to the Secretary on methods of maximizing opportunities for beginning farmers and ranchers. Members include representatives of FSA; the Cooperative State Research, Education, and Extension Service (CSREES); State beginning farmer programs; commercial lenders; private nonprofit organizations with active beginning farmer and rancher programs; community colleges or other educational institutions with demonstrated experience in training beginning farmers and ranchers; other entities or persons providing lending or technical assistance to beginning farmers and ranchers; and farmers and ranchers.

The 2002 Farm Bill built on the Agricultural Credit Improvement Act of 1992 by authorizing the Secretary to provide higher payments to beginning farmers and ranchers in some of USDA's conservation programs. It also authorized the Secretary to provide incentives to beginning farmers and ranchers to participate in conservation programs to foster new farming and ranching opportunities and enhance environmental stewardship over the long term. Since FY 2003 through April 2007, the Natural Resources Conservation Service (NRCS) has approved $292 million in Environmental Quality Incentives Program (EQIP) contracts for beginning farmers and ranchers and $134 million for limited resource farmers. In FY 2005 and 2006, NRCS invested $6.5 million and $10 million, respectively, under EQIP for a special initiative to assist beginning and limited resource farmers and ranchers to implement conservation practices on their land. NRCS has also encouraged their State offices to give extra points in their Farm and Ranch Lands Protection Program ranking criteria for farms with succession plans.

The 2002 Farm Bill also authorized the Secretary to establish a Beginning Farmer and Rancher Development Program to provide training, education, outreach, and technical assistance initiatives for beginning farmers and ranchers. No funds have been allocated to implement this program. However, CSREES provides grants to organizations that assist beginning farmers and ranchers through the (1) Outreach and Assistance for Socially Disadvantaged Farmers and Ranchers Program (Section 2501) and (2) National Research Initiative Grants Program (Agricultural Prosperity for Small and Medium-Sized Farms). The 2002 Farm Bill authorized $25 million in annual appropriations for FY 2002 through 2007.

New entrants into agriculture can participate in price and income support programs and crop insurance on the same terms as other producers. In addition, the Risk Management Agency (RMA) provides partnerships to organizations to assist beginning farmers and ranchers in risk management. Following the recommendations of the Advisory Committee, RMA has targeted partnership applicants that were sensitive to the needs of beginning farmers and ranchers and those that planned to partner with organizations that assist beginning farmers and ranchers. Seven organizations assisting beginning farmers and ranchers received partnerships in the first solicitation in 2003. Except for 2007, RMA has continued to include beginning farmer and rancher language in their annual partnership request for applications, including the Community Outreach and Assistance Partnership Program Announcement published March 1, 2006. On June 16, 2006, RMA also announced a rural initiative for new farmers, making available $500,000 to fund educational programs to teach refugees and other low-income individuals who produce specialty crops about risk management and good business practices.

In 1999, USDA implemented Departmental Regulation (DR) 9700-1, "Small Farms Policy". The Advisory Committee recommended USDA develop and implement a mission focus to heighten awareness and coordinate beginning farmer and rancher opportunities, similar to that established in DR 9700-1 for small farms. On August 3, 2006, DR 9700-1 was amended to become a "Small Farms and Beginning Farmers and Ranchers Policy." This has resulted in a Small Farms and Beginning Farmers and Ranchers Council, which I as Deputy Secretary, Chair. DR 9700-1 also requires beginning farmers and ranchers policy to be reflected in all USDA mission area and agency statements, strategic plans, performance plans, and performance goals.

2007 Farm Bill

The Administration is recommending a broad package of proposed changes to several Farm Bill titles to provide additional support to beginning farmers and ranchers.

As part of the commodity title, the Administration is recommending to provide beginning farmers a 20-percent increase in their direct payment rate, adding $250 million to producer income over 10 years. After the initial five years, these producers would no longer be eligible for the higher direct payment rate. This change would better prepare beginning farmers to face the initial financial burdens associated with entering production agriculture.

Under the conservation title, we are recommending that 10 percent of farm bill conservation financial assistance be reserved for beginning farmers as well as socially disadvantaged producers under a new Conservation Access Initiative. This new initiative would maintain the higher rates of Federal cost-share, but also direct a greater technical assistance percentage than the traditional program to better address the needs of socially disadvantaged agricultural producers. Funds set-aside under the Conservation Access Initiative within the Conservation Innovation Grants would be used for technology transfer, farmer-to-farmer workshops, and demonstrations of conservation success to further adoption of innovative conservation practices.

As part of the credit title, we are recommending to enhance the existing Beginning Farmer and Rancher Downpayment Loan Program to help ensure the success of the next generation of production agriculture by cutting the interest rate in half to two percent; deferring the initial payment for one year; eliminating the $250,000 cap on the value of property that may be acquired by a beginning farmer wishing to obtain a down payment loan and replacing it with a maximum down payment loan amount of $200,000; decreasing the minimum beginning farmer contribution from 10 percent of the property purchase price to 5 percent; and adding socially disadvantaged (SDA) farmers and ranchers as eligible applicants for this program

We are also recommending a doubling of the statutory target for the percentage of USDA direct operating loans (OL) that will be prioritized to assist beginning and SDA farmers from 35 percent to 70 percent; prioritizing 100 percent rather than 70 percent of direct farm ownership (FO) loans to first meet the needs of beginning and SDA farmers; and overhauling the federal repooling procedures to ensure that these targets are reserved only for beginning and SDA farmers to the maximum extent possible.

The Administration also recommends increasing the existing statutory limits of $200,000 for direct farm ownership loans and $200,000 for direct operating loans to a maximum of $500,000 indebtedness for any combination of the two loan types. The higher, combined limit will allow a better matching of loan type, amount, and purpose to an individual applicant's credit needs. It will also improve access to capital, and therefore the competitiveness, of beginning farmers and other FSA loan applicants.

CONCLUSION

Market incentives augmented by Federal and State programs will determine the future supply of farmers and ranchers. If farm production provides a sufficient return, capital investment and people will enter production agriculture. Low rates of return will discourage investment and cause farm failures. Competition will drive successful producers to adopt technology and achieve efficient operations, while inefficient producers will exit agriculture.

The operation of these market forces is critically important for growth, productivity gains, and ensuring affordable food for Americans.

Nevertheless, there are a number of issues concerning the adequacy of the future workforce needed to farm the nation's agricultural lands. These issues include the potential divergence between the level of younger, new farm entrants into production agriculture and the exit of older retiring farmers; the potential barrier to entry for new farmers created by rising farmland values and their effect on the capital costs of acquiring an efficient farm operation; the rising complexity of farm production caused by changing markets, globalization, new technologies, economies of scale, environmental concerns, and other factors.

Uncertainty over the outcome of these issues and the structural changes that have accompanied the growth in agriculture have led to programs and policies that aid new farm formation or the successful succession of farms from one generation to the next. The 2007 Farm Bill provides an excellent opportunity to improve the effectiveness of these efforts.

That completes my statement, Mr. Chairman.

In: America's Family Farms
Editor: Efren J. Tamayo

ISBN: 978-1-60741-751-4

Chapter 3

AGING FARMERS HEARING- GODWIN TESTIMONY

Oregon State University

June 18, 2007

To: Special Committee on Aging, United States Senate

From: Derek Godwin, Staff Chair in Marion County and Watershed Management Specialist

Re: Testimony for Special Committee on Aging

The following testimony is provided by Derek Godwin and includes information gathered from John Burt, professor emeritus at OSU, and Mark Green, professor at OSU and Director of the Austin Family Business Program.

Farmers face many challenges like making daily decisions on when to plow, till, plant, fertilize, and harvest, or management decisions involving their sizeable assets, numerous employees, a multitude of rules and regulations and a market place that changes every day. In addition to these challenges, competition from foreign markets, environmental concerns, low prices and many global factors also make the future uncertain.

These issues may be similar to a lot of other businesses, but what makes farming unique is the family nature of the business. Probably 95% of commercial agriculture is structured as family owned businesses - partnerships, corporations or single proprietorships - the basic underlying structure making agriculture so productive over the last century. Husbands and wives, brothers and sisters, parents and their children working side by side on land passed on over several generations.

In the last two U.S. Census of Agriculture reports the average age of farm owners continues to increase towards 60 years of age. This means that our communities and society in general can anticipate an unprecedented transfer in ownership of land-based business over the next couple of decades. In addition, the value of agriculture property has appreciated significantly over the years which adds complexity to how and if farms will transition to the next generation. We are at a critical stage in planning for the future of agriculture: recruiting and training the next generation of farmers and ensuring farms will continue to be viable, healthy operations. Every family owned business has to deal with transitions, but it seems to be reaching crisis proportions in agriculture.

The following paragraphs describe typical family and transition management issues with farming families in Oregon and other states. This testimony highlights these issues to assist the Committee on Aging in understanding the issues and explore possible solutions in supporting a healthy, vibrant future for agriculture and our families in the United States.

So what is preventing young people from staying on the farm? First and foremost is profitability. Decisionmakers and consumers alike are stuck on "cheap food" as a goal both in the local super market and as a way to compete in foreign markets. If we don't get this one solved, there won't be a future in farming in America for the next generation. In Oregon, one recent marketing tactic that is gaining success is educating people to "buy local" and support local farms. This education is helping increase demand at grocery stores for locally grown products even if they are more expensive than foreign and out-of-state products. Another recent success in Oregon is to increase the amount of locally grown products being used in state schools. Schools have historically held contracts with companies that used foods from other states and countries solely because they were the cheapest to obtain.

Next is the difficult access into farming. In reality, you are either born or marry into it. Small farms are an option, but small farms have a hard time surviving and they are often highly leveraged and subsidized by off farm income. Farmers markets, roadside stands, U-Pick farms and community farming all find a hard way to gain acceptance in the market place, but are some of the best ways for small farms to get started.

Estate taxes are another major burden facing the inter-generation transfer of a farming operation. A farm may be asset heavy whether it is profitable or not, and the last transfer between the parent and offspring can still be heavily taxed. So much so that heirs often go into debt and/or sell parts of the property to pay the tax bill.

The above are the more tangible barriers to choosing farming for a career. How about long hours, low pay, isolation, working with your parents and few vacations or weekends off? To paraphrase a local farmer: How are you going to keep them on the farm after they have seen what a job at Microsoft pays?

All of this isn't new or necessarily limited to farming; but, it's a critical time to explore solutions to resolve these issues. Here are a few ideas from my own point of view.

On the big picture scale, consider eliminating or seriously modifying the estate tax for family owned businesses in order to allow for the successful transition among generations.

Look for ways to make access into farming a reality for more young people. This could range from low interest business loans to scholarships for people attending university and technical schools to study agriculture and business with the intent on returning to the family farm.

Encourage agri-business, agricultural organizations and non-profits and even local farmers to take a more direct role in nurturing those young people who want to get into or remain in agriculture. They are as close as your local FFA Chapter, 4-H Club, Community College or University. They should be available for internships, work-study programs or job shadowing. The more young people are exposed to agriculture as a career, the more likely they will opt for it as a career choice and eventually find their way home.

Finally, create local support and advisory groups for farm families facing these challenges. Look for networks that link schools, non-profit and other organizations together with farm families.

If we start with the premise that keeping the "family" in family farms is a good idea - and it seems that the great success in America's agriculture over the years is because its structure is based on a family-owned business - future success will come because we have supported families on farms and their ability to effectively transition to the next generation.

Oregon State University Extension Service and the Austin Family Business Program at Oregon State University have been developing and delivering educational assistance to help farmers and woodlot owners create effective succession plans. These programs provide information about the legal and economic aspects of managing and transferring family farms and woodlands; but more importantly, they provide steps within the complicated emotional context of family ties. Oregon and colleagues in many other states have found that the most critical component in creating a successful transition plan is the ability for the family to effectively communicate. These Extension programs incorporate communication techniques while guiding landowners through the following steps in creating a succession plan:

1. Discuss and write down your goals (vision) for the property and the family.
2. Discuss these written goals with your family
3. Create a family business entity to own the land
4. Have regular family meetings to discuss the business and share your passion
5. Set family employment policies before you hire any family members
6. Discuss and write down important decisions
7. Create non-financial reasons for the family to keep the property
8. Get your kids and grandkids out to work and have fun on the property
9. Create a governance structure that will survive your passing
10. Remember to have fun!

No matter what the future holds for agriculture, whether it is a move towards organic markets, niche markets, new products, etc., success will come because we have worked hard to generate solutions that effectively support families on farms. I commend this Committee on Aging for their foresight in considering these issues, and I hope this testimony has provided some ideas on possible solutions.

In: America's Family Farms
Editor: Efren J. Tamayo
ISBN: 978-1-60741-751-4

Chapter 4

AGING FARMERS HEARING- KERSHAW TESTIMONY*

Isaac Kershaw

Good morning. Chairman Kohl, Ranking Member Smith and members of the Special Committee on Aging. My name is Isaac Kershaw and I appreciate the opportunity to address the committee on behalf of the National FFA Association and the National Council for Agricultural Education. Today, I'd like to share with you our insight regarding issues associated with aging farmers and ranchers and the challenges facing young people who have a desire to enter production agriculture.

The retirement of aging baby boomers and the graying of American farmers rightly raise significant questions: Who will farm the land? Will we have enough young people willing to enter the field in the future? Will they be successful? Who will prepare them?

Those familiar with agricultural education know there are currently more than a million students in our public schools preparing for careers in agriculture, food and natural resource industry areas. A half million of them are members of the FFA, an organization dedicated to developing the potential of students for leadership, personal growth and career success as an integral part of agricultural education. A rising FFA membership is at the highest point in 28 years and suggests increasing interest by students in agricultural careers. Some 12,000 teachers in 7,200 schools work with these students every day, helping them discover their talents, explore opportunities in agriculture and channel their efforts for career preparation. It's a system that has served generations of the nation's farmers and ranchers, and today it is bringing the best and brightest to agriculture.

Agricultural education programs are designed to create awareness of opportunities in agriculture, motivate students to begin their preparation for a lifelong career in agriculture, and prepare them with the knowledge and skills necessary for success in higher education and on the job. Students explore careers and begin to understand that 21st century agriculture is a global enterprise based in science, which demands continued growth in discovery and application. Many of these young students choose to pursue careers in farming and ranching. Many others take a different path to an agricultural career, but all are related to the science,

---

* This is an edited, reformatted and augmented edition of statements made before the Senate Committee, Special Committee on Aging, by Isaac Kershaw on June 21, 2007

business and technology of agriculture. All are tied to food, fiber, fuel and natural resource management.

Plant and animal sciences serve as the foundation body of knowledge for these programs. Students will understand the marketing, processing and distribution systems that support production. Resource management is taught through agricultural education and is critical to success in establishing and managing the business of a complex enterprise. Individuals who plan for a career in production agriculture need strong academic knowledge and skills. Quality agricultural education programs integrate science, math, communication, business and technology into the curriculum and treat these academic competencies as critical for success in college and on the job.

I submit we have an ideal system in place to produce the results we seek. Agricultural education and FFA have the ability to attract, motivate and help prepare the next generation of agricultural leaders, managers, scientists and producers. But we will best serve the interests of agriculture and the nation if we dramatically expand the reach of this program to more of the nation's high schools.

The agricultural education community is in the process of implementing a plan of action designed to significantly increase the number and quality of agricultural science programs in this country. Working under the direction of the National Council for Agricultural Education, we have adopted a long-range goal to increase the number of programs from 7,200 today to 10,000 by the year 2015. This ambitious effort is viewed as a key strategy for attracting the talent and commitment of those who will keep American agriculture thriving and productive.

We know that significant financial resources are necessary for young men and women to enter careers in production agriculture. It will require a comparable investment in education for them to acquire the knowledge tools necessary to be successful. This should be seen as a great and necessary investment in our national future.

The focus of the Special Committee on Aging, to facilitate an effective transition to a new generation of agricultural producers, is timely and critical. The more students we get through the educational pipeline, the greater the number prepared to successfully fill the shoes of the farmers and ranchers who precede them. But we need help with this endeavor. We need more research. We need program innovation. We need more investment through education and agriculture appropriations. We need language in the Farm Bill and in educational policy that strengthens the role of agricultural education and FFA in securing the future agriculture base of the nation. There is a real opportunity for the Federal Government to take an active role as we move forward with the initiatives of having 10,000 quality programs by 2015.

Mr. Chairman and members of the committee, thank you for the opportunity to be here today on behalf of the National FFA Association and the National Council for Agricultural Education to express our views. I appreciate your public service and wish you success in your deliberations.

In: America's Family Farms
Editor: Efren J. Tamayo

ISBN: 978-1-60741-751-4

Chapter 5

AMERICA'S DIVERSE FAMILY FARMS*

United States Department of Agriculture

American farms encompass a wide range of sizes, ownership structures, and business types. A farm classification developed by USDA's Economic Research Service (ERS) categorizes farms into more homogeneous groupings for reporting and evaluation purposes. The typology is based largely on farm sales, organization, and the farm operator's primary occupation.

FARM TYPES FOR 2004

The farm classification developed by ERS focuses on the "family farm," or any farm organized as a sole proprietorship, partnership, or family corporation. Family farms exclude farms organized as nonfamily corporations or cooperatives and farms with hired managers.

Small Family Farms (sales less than $250,000)

Limited-resource. Farms with gross sales less than $100,000 in 2003 and less than $105,000 in 2004.[1] Operators of limited-resource farms must also receive low household income in both 2003 and 2004. Household income is considered low in a given year if it is less than the poverty level for a family of four, or it is less than half the county median household income. Operators may report any major occupation except hired manager.

Retirement. Small farms whose operators report they are retired (excludes limited-resource farms operated by retired farmers).

* This is an edited, reformatted and augmented version of a U. S. Department of Agriculture publication dated June 2007.

Residential/lifestyle. Small farms whose operators report a major occupation other than farming (excludes limited-resource farms with operators who report nonfarm work as their major occupation).

Farming-occupation. Farms whose operators report farming as their major occupation (excludes limited-resource farms whose operators report farming as their major occupation).

Low-sales. Gross sales of less than $100,000.
Medium-sales. Gross sales between $100,000 and $249,999.

Large-Scale Family Farms (sales of $250,000 or more)

Large family farms. Farms with sales between $250,000 and $499,999.
Very large family farms. Farms with sales of $500,000 or more.

Nonfamily Farms

Nonfamily farms. Farms organized as nonfamily corporations and cooperatives, as well as farms operated by hired managers. Also includes farms held in estates or trusts.

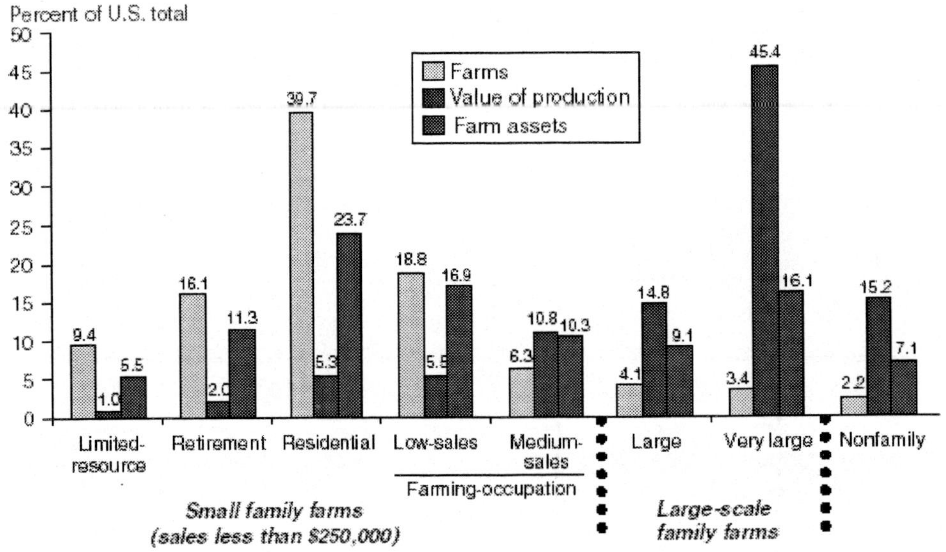

Distribution of farms, value of production, and farm assets, 2004

FARMS, ASSETS, AND PRODUCTION

Most farms are small, and most farm assets are on small farms, but small farms account for only 25 percent of production.

Ninety percent of farms are small, and these farms account for 68 percent of farm assets, including 61 percent of the land owned by farms.

Large-scale family farms and nonfamily farms account for 75 percent of production.

Family farms of different sizes account for 98 percent of farms and 85 percent of production.

Differences among farm types are illustrated in this brochure using 2004 data from the Agricultural Resource Management Survey, an annual survey conducted by USDA's Economic Research Service (ERS) and National Agricultural Statistics Service (NASS). Data from various censuses of agriculture are used to follow trends over time.

SHIFTS IN PRODUCTION TO MILLION-DOLLAR FARMS

Sales of farm products have shifted to million-dollar farms (sales of $1,000,000 or more, in 2002 dollars) over the past two decades

Million-dollar farms—less than 2 percent of all farms—accounted for 48 percent of the sales of farm products in 2002, up from 23 percent in 1982.

At the same time, the share of sales by farms with less than $250,000 fell from 53 percent to 24 percent.

The most common specializations among million-dollar farms are specialty crops (27 percent of the group), poultry (19 percent), dairy (14 percent), and hogs (10 percent).

Most million-dollar farms (88 percent) are family operations. Less than 1 percent of million-dollar farms are operated by large, publicly held corporations.

For more information about the diversity of U.S. farms, see ERS's **Structure and Finances of U.S. Farms: Family Farm Report, 2007 Edition.**

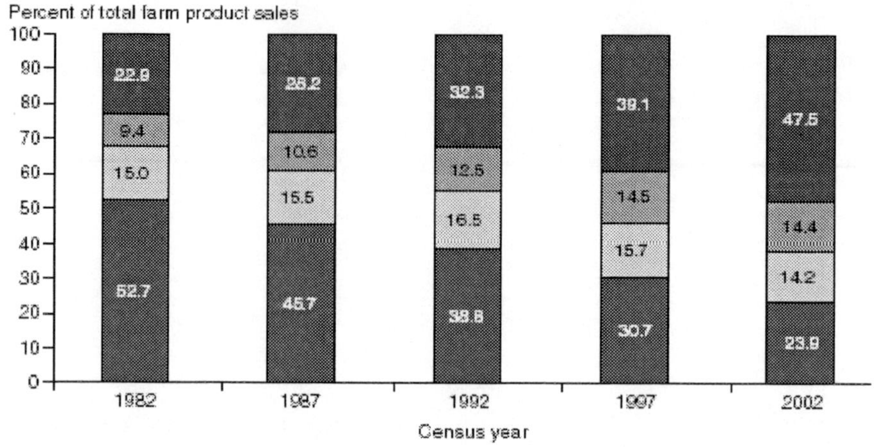

Note: Sales are expressed in constant 2002 dollars, using the Producer Price Index for farm products to adjust for price changes.

Farm product sales by constant-dollar sales class (2002 dollars), 1982 to 2002

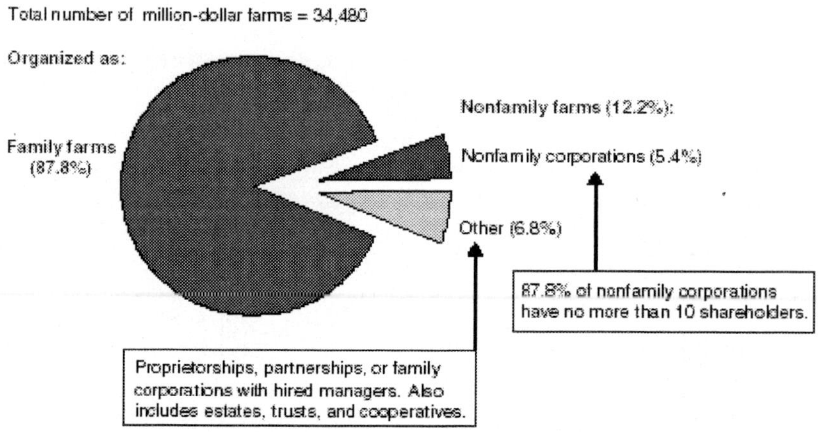

Organization of farms with gross sales of $1 million or more, 2004

INCREASE IN TYPICAL ENTERPRISE SIZE

The shift of sales to larger sales classes is reflected by an increase in typical enterprise size over time.

- Between the 1987 and 2002 Censuses of Agriculture, typical acres harvested roughly doubled for most types of vegetables, as well as for field crops.
- We use animals to define typical livestock enterprise sizes. In 1987, half of all hog sales came from farms selling 1,200 or fewer hogs, and half from farms selling more. By 2002, that typical enterprise size rose to 23,400 hogs.
- Typical fattened cattle and dairy enterprises also grew in size, approximately doubling and tripling (respectively) over the 15-year period.

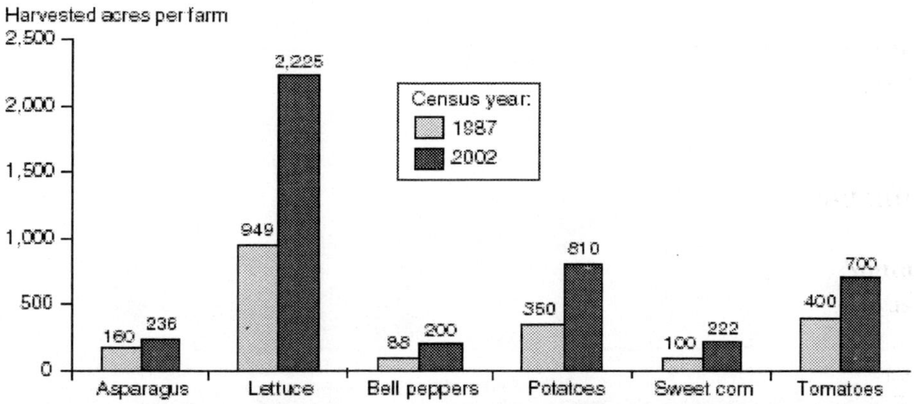

Note: "Typical enterprise size" is the median acres harvested of a particular vegetable. Half of all harvested acres of a vegetable came from farms harvesting more than the typical acreage, and half came from farms harvesting less. Enterprise size differs from farm size because a farm may have multiple crop or livestock enterprises.

Change in typical enterprise size for selected vegetables,1987 to 2002

FARM FINANCIAL PERFORMANCE

Small farms are more likely to have a negative profit margin, but some are profitable.

Nearly half or more of the farms in each small farm type had a negative operating profit margin in 2004.

Other small farms were more profitable—between 15 and 28 percent of each small farm group had an operating profit margin of at least 20 percent.

Nevertheless, an even greater share of large-scale farms had operating profit margins that high, and most large-scale farms had positive margins.

Only 3 percent of U.S. farms are classified as vulnerable (negative net farm income and debt/asset ratio greater than 40 percent), and 56 percent of these farms are residential/lifestyle.

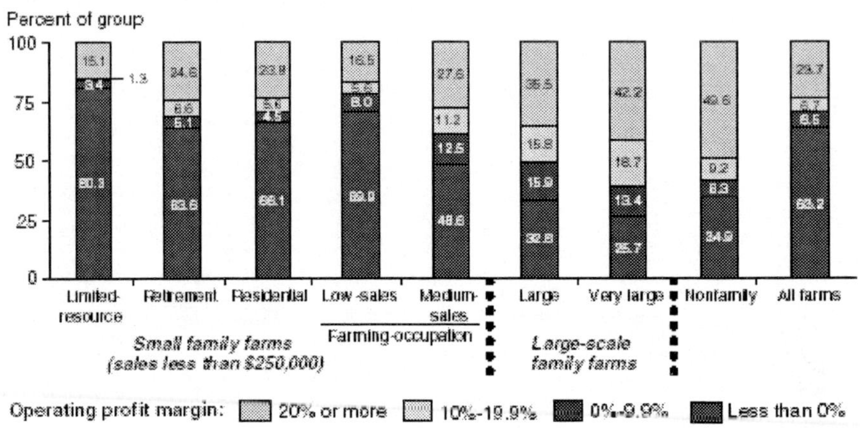

Note: Operating profit margin = 100% X (net farm income + interest paid - charge for unpaid operators' labor and management)/gross farm income.

Farms by operating profit margin, 2004

FARM OPERATOR HOUSEHOLD INCOME

Most types of farm households have a median income at or above the median for all U.S. households, when both farm and off-farm sources of income are considered.

Only two types of farm households—those operating limited-resource and low-sales farms—have a median household income substantially below the median for all U.S. households.

Operators of small farms (especially if farming is not their major occupation) often report losses from farming.

Off-farm work provides the primary source of income for most types of farm households.

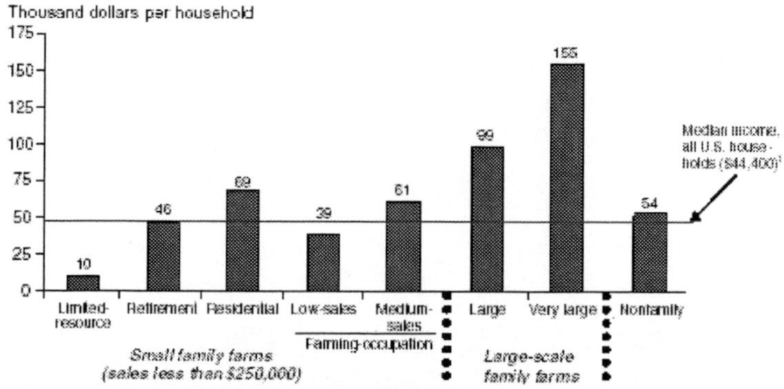

Note: Median income falls at the midpoint of the distribution of households by income, dividing households into two groups. Half of the households have income above the median, while the other half have income below that level. Farm household income is not estimated for nonfamily farms.

[1]Median income for all U.S. households is from the U.S. Census Bureau, 2005 Current Population Survey.

Median income of operator households, 2004

Average (mean) farm operator household income by source, 2004

Type of farm operated	Total average income	Income from farming		From off-farm sources		
		Amount	Negative	Total	Earned[1]	Unearned[1]
	Dollars per household		*Percent of households*	*Dollars per household*		
Small family farms:						
Limited-resource	7,680	-5,902	72.1	13,582	3,463	10,118
Retirement	62,468	4,128	50.8	58,339	20,252	38,087
Residential/lifestyle	96,515	-365	64.4	96,879	83,548	13,331
Farming-occupation:						
Low-sales	63,043	4,925	44.4	58,118	36,950	21,168
Medium-sales	70,365	34,354	24.6	36,011	26,241	9,769
Large-scale family farms:						
Large family farms	125,120	80,250	16.8	44,870	33,238	11,633
Very large family farms	272,527	225,094	16.3	47,434	29,320	18,114
Large-scale family farms	81,596	14,317	52.8	67,279	48,818	18,461

Note: Farm household income is not estimated for nonfamily farms.

[1]Earned income comes from off-farm self-employment or wage or salary jobs. Unearned income includes interest and ividends, benefits from Social Security and other public programs, alimony, annuities, net income of estates or trusts, private pensions, etc.

Government Payments

Government payments go primarily to family farms, but commodity-related payments tend to go to large-scale family farms, while conservation payments tend to go to small family farms.

Commodity program payments reflect acreage in crops historically eligible for support; 59 percent went to family farms with at least $250,000 in sales in 2004.

Conservation and commodity payments are distributed among different farms; 80 percent of conservation payments went to family farms with less than $250,000 in sales.

Because commodity-related payments make up 83 percent of all government payments, the distribution of total government payments and commodityrelated payments is similar.

The share of farms receiving government payments is particularly high for medium-sales small farms and large-scale farms, largely due to their participation in commodity-related programs.

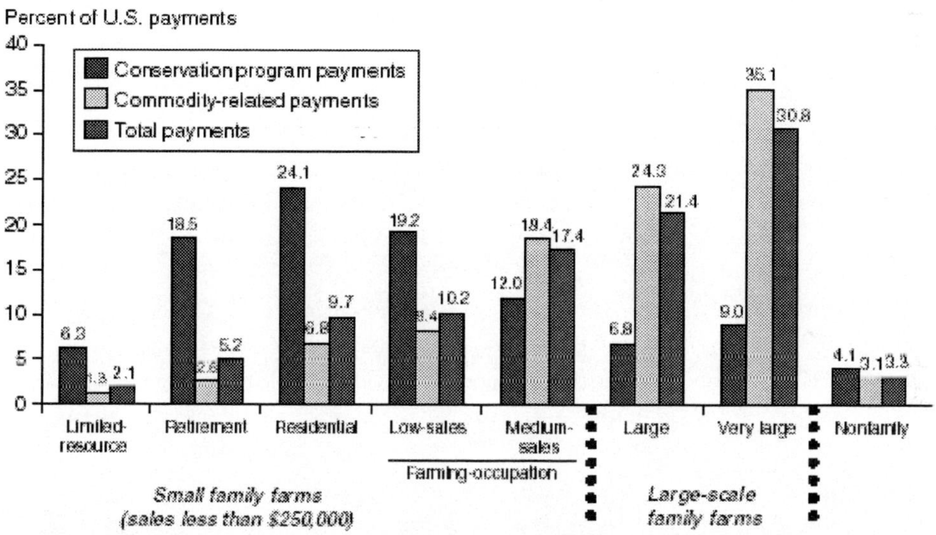

Conservation programs: Conservation Reserve Program (CRP), Wetlands Reserve Program (WRP), and Environmental Quality Incentives Program (EQIP) payments.

Commodity-related programs: Direct payments, countercyclical payments, loan deficiency payments, marketing loan gains, net value of commodity certificates, peanut quota buyout, milk income loss contract payments, agricultural disaster payments, etc.

Distribution of government payments by type of payment, 2004

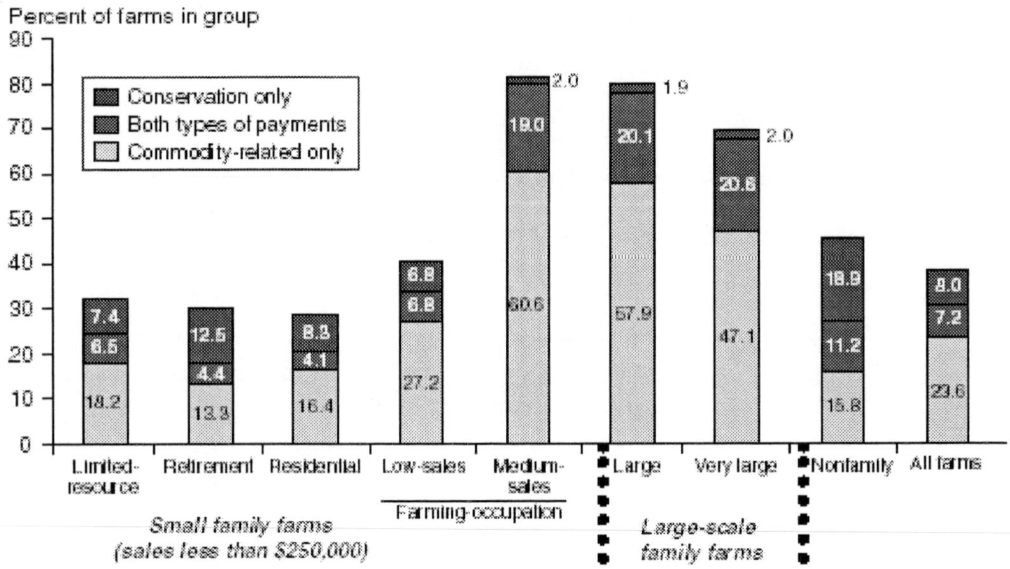

Farms receiving government payments by type of payment, 2004

SOURCES OF DATA ON GOVERNMENT PAYMENTS

The Agricultural Resource Management Survey (ARMS) relies entirely on responding farm operators for program-related information and records only the payments that they report receiving. In contrast, administrative data—based on payment records kept by the agencies involved—also include payments received by people who do not farm, mainly nonoperator landlords. Thus, the survey shows different levels and composition of government payments than do administrative data. The advantage to ARMS is that it includes detailed information about the farms and operators receiving the payments. Administrative data contain limited information about recipients of government payments.

Farm Policy and Family Farms

Sales of farm products shifted dramatically to farms with sales of $1,000,000 or more between the 1982 and 2002 Censuses of Agriculture. Most of these million-dollar farms are organized as family farms, and those organized as nonfamily corporations generally have no more than 10 stockholders.

Farming is still an industry of family businesses. Ninety-eight percent of farms are family farms, and they account for 85 percent of farm production.

Different farm policies affect distinctly different sets of farmers. Commodity program payments largely flow to large and very large family farms, and are shifting toward such farms along with production. Policies addressing natural resource and conservation concerns, which are often based on land ownership, have stronger impacts on smaller family farms.

The nonfarm economy is critically important to operators of small family farms. Because small-farm households rely on off-farm work for most of their income, general economic policies, such as tax or economic development policy, can be as important to them as traditional farm policy.

End Note

[1]The original gross sales cutoff was established at $100,000 for 2003. The cutoff for subsequent years is adjusted by the index of prices paid by farmers.

In: America's Family Farms
Editor: Efren J. Tamayo
ISBN: 978-1-60741-751-4

Chapter 6

ECONOMIC WELL-BEING OF FARM HOUSEHOLDS*

Carol A. Jones, Hisham El-Osta and Robert Green

Farm subsidy programs in the 1930s were largely prompted by concern for the chronically low, and highly variable, incomes of U.S. farm households. Seventy years later, commodity-based support programs are still prominent, even though the income and wealth of the average farm household now exceed those of the average nonfarm household—wealth by a large margin.

Farm households continue to face variability in income due to weather and natural disasters. Household income is most variable for the small segment that operates commercial farms (above $250,000 in annual sales). Relative to small farms, these farms achieve greater economies of scale, generate higher profit margins, and their households realize a larger share of their income from farming. However, the substantial net worth of these households acts as a cushion against uncertain farm income, much as off-farm income does for households operating smaller farms.

In a variable-income/high-wealth sector such as farming, economic well-being measures based on both income and wealth can provide a better signal of household capacity to support a consistent living standard than income measures alone. In 2003, 5 percent of farm households had both income and wealth below the respective U.S. household medians, and those households, on average, spent more on basic consumption than they earned in income. Households with low income and low wealth are less likely to receive farm payments, excluding conservation programs; by contrast, only 3 percent of households receiving payments had income and wealth below the U.S. household medians for each.

* This is an edited, reformatted and augmented version of a U. S. Department of Agriculture publication dated March 2006.

INCREASING FARM HOUSEHOLD PARTICIPATION IN OFF-FARM EMPLOYMENT AND INVESTMENT IS KEY TO WELL-BEING

The average income of farm households increased from half of nonfarm household income (per capita) in the 1930s to relative parity by the 1970s. In every year since 1996, average income for farm households has exceeded the average U.S. household income by 5 to 17 percent. Today, the economic portfolios of most farm operator households are highly diversified. Off-farm sources of income (including employment earnings, other business activities, other investments, and transfer payments) provided 85-95 percent of household income over 1999-2003, up from around 50 percent in 1960.

Operators of family farms in all sales classes had average household income exceeding the 2003 U.S. average for all households ($59,083). However, farm households are following diverse paths to economic wellbeing. Commercial farms (annual sales above $250,000, representing about 7 percent of U.S. farms) produce about 70 percent of total farm sales and have an average operating profit margin[1] greater than 10 percent, with economic performance and farm share of household income increasing with farm size within the commercial segment (see Economic Brief No. 6, *Growing Farm Size and the Distribution of Farm Payments*). Very large commercial farms (sales greater than $500,000) average household income about four times the U.S. household average. Though farm income provided 80 percent of household income for the average very large commercial farm operator in 2003, off-farm income still contributed around $44,033 per year.

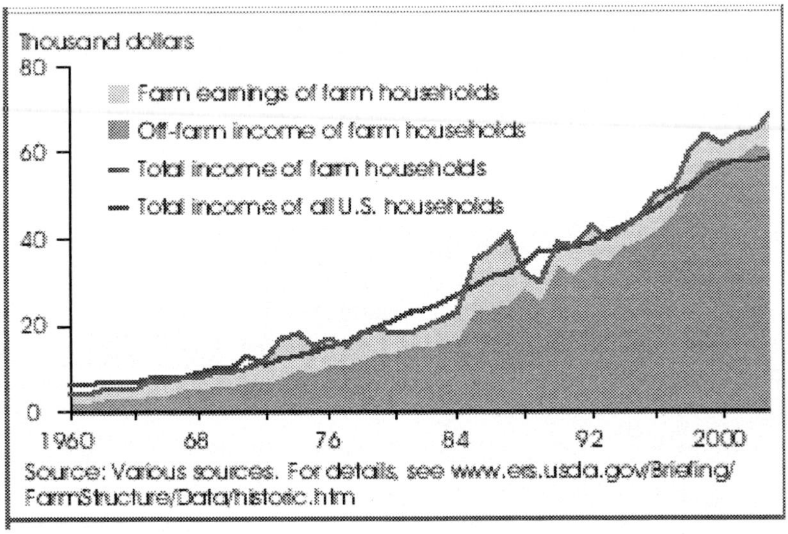

Figure 1. Average household income for farm households compared with all U.S. households, 1960-2003

Across all other sales classes, farms have negative farm operating profits, on average, and their households draw most of their income from off-farm sources. Farm operating profit margins become more negative and shares of household income from farm sources decrease as farm size diminishes. Households just below commercial farms ($100,000-$249,999 in sales) represent 8 percent of U.S. farms and produce 17 percent of sales. Their average household income was $67,275 in 2003. The remaining 85 percent of farms produce around 15 percent of sales, and earn negligible income from farming. The operators of these smaller farms (particularly those with less than $10,000 in sales) disproportionately identify their primary occupation as "other than farming or ranching" or as "retired." The "other occupation" group, who operate 42 percent of all farms, is more integrated into the off-farm economy than the farmer/rancher or the retired, and relies primarily on earned income (off-farm wages and salaries and off-farm business income.)

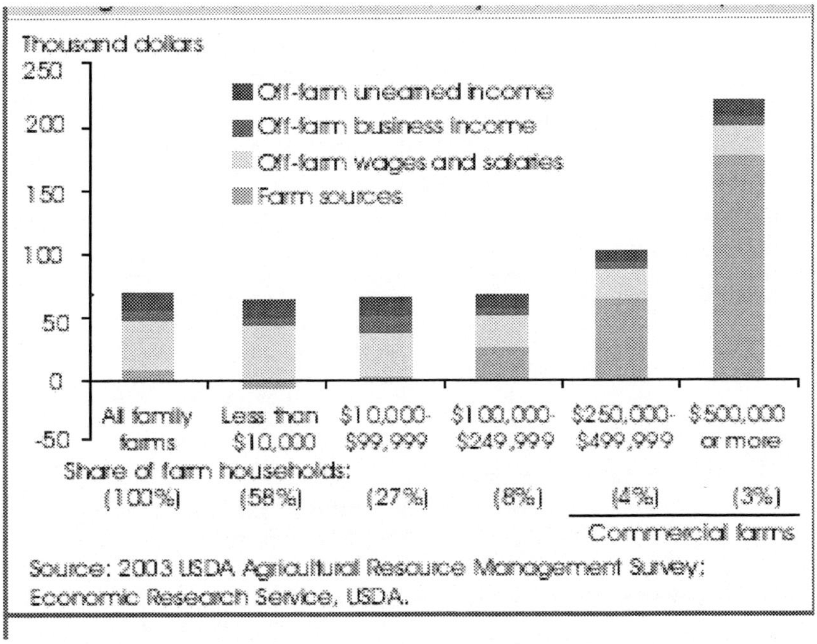

Figure 2. Average household income varies by sales class of farm, 2003

FARM HOUSEHOLD INCOME IS MOST VARIABLE FOR HOUSEHOLDS WITH HIGHEST NET WORTH

Farm households as a group no longer experience chronically low incomes relative to nonfarm households. On the other hand, farm households do continue to experience more variable income from year to year. However, it is the 7 percent of farm households operating commercial farms, who derive a majority of household income from the farm, that experience the greatest degree of variability in household income from year to year. The 8 percent of farm households operating the next size class of farms ($100,000-$249,999) also experience variability in household income, though the effect is dampened because about two-thirds of their income comes from offfarm activities.

Distribution of Income and Wealth Across Farm Households

Within a given year, the variability of income across farm households tends to mirror that for all U.S households. However, farm households tend to have lower incomes at the low end of the income spectrum than nonfarm households, and higher incomes at the high end. The share of farm households with negative household income was 6 percent in 2003, versus 1 percent of all U.S. households.

In contrast to all U.S. households, where wealth is highly concentrated at the top end of the distribution, wealth is more equally distributed across farm households. Nonetheless, differences exist by farm size and by age/retirement status. Across size classes, the variation in farm household wealth roughly mirrors the variation in income levels: farm net worth and, to some extent, nonfarm net worth are higher for households operating larger farms. So the larger farms can counter their greater exposure to variable income with higher net worth – on average in excess of $1 million, and closer to $2 million for very large commercial farms.

Again, the households operating larger farms (sales greater than $100,000) are the ones more likely to experience the effects on household income from variability in farm income: in 2003, around 13 percent of households operating larger farms had negative household income, compared with 4 percent of households operating smaller farms (figure 3). In contrast, the likelihood of incurring losses from farm operations is highest among the smaller farms (less than $100,000 in sales). But these latter households acquire virtually all of their income from off-farm sources, so negative farm earnings seldom translate into negative household income.

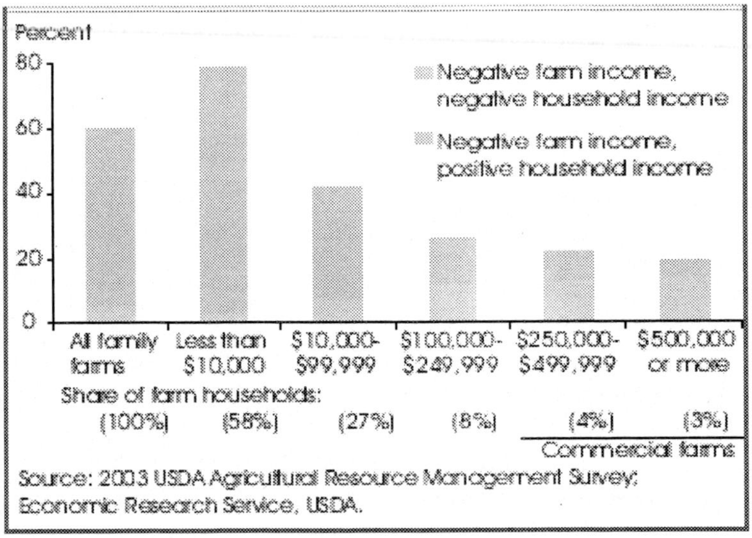

Figure 3. Share of farm households with negative farm income, by whether total household income is positive or negative, and by sales class, 2003

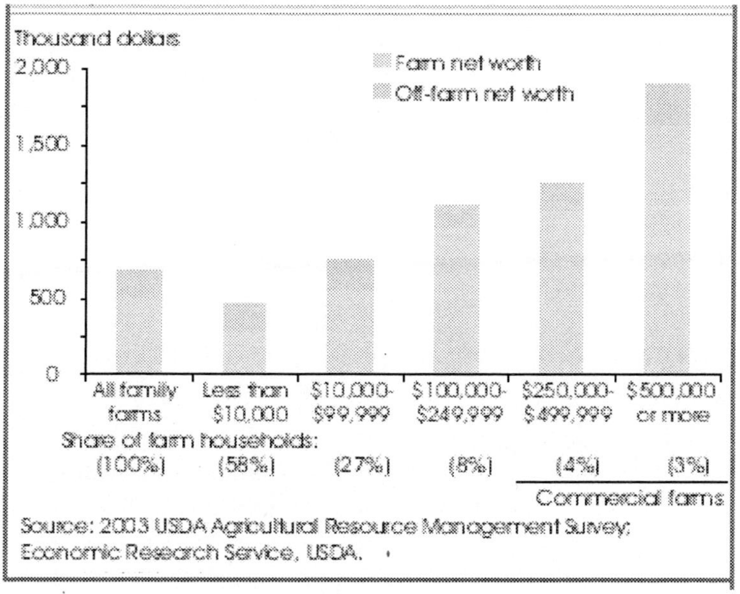

Figure 4. Average farm household net worth, by sales class, 2003

A JOINT INCOME-WEALTH INDICATOR IS MORE INDICATIVE OF FARM HOUSEHOLD WELL-BEING

In 2003, median wealth of farm households ($416,250) was five times the estimated median wealth of all U.S. households ($89,578). (By definition, 50 percent of households have wealth lower than the median—also known as the 50th percentile—of wealth). Seventy-three percent of farm household net worth is in farm equity (plus an unknown share is in nonfarm business equity), whereas 17 percent of the net worth of U.S. households is in business equity. Farmland, which has appreciated greatly in recent years, particularly near urban centers, currently represents about 60 percent of farm household wealth. Excluding farm wealth, median nonfarm wealth of farm households ($83,750) was almost as high as estimated median total wealth of all U.S. households.

Gauging what share of farm households has low economic wellbeing is challenging, because farming is characterized by variable income but also by high wealth. During periods of low income, farm households may be able to maintain living standards by borrowing against, or liquidating, assets. Consequently, household income for an individual year, the standard measure of economic well-being, is not necessarily a good indicator of a farm household's ability to support a given consumption level through time. And wealth is particularly important for the retired and near-retired, who may be drawing down wealth accumulated over their lifetime, rather than spending current income, to support their standard of living.

To create a well-being indicator that accounts for both income and wealth, we separate farm households by low and high income and low and high wealth, using the U.S. household medians for income and wealth as the dividing lines. By definition, 50 percent of U.S. households had income greater than the U.S. median income ($43,378) in 2003. In contrast,

54 percent of farm households had income greater than that level in 2003. However, the big difference is in the distribution of wealth across the groups: 92 percent of all farm households—in contrast to 50 percent of all U.S. households—had wealth greater than the U.S. median (estimated to be $89,578 in 2003).

So who is in the small group of low-wealth households? On average, the low-wealth group was younger (virtually none was retired), operated substantially fewer acres, and generated lower farm sales than the farm operator population as a whole. They reported substantial losses in the off-farm component of household wealth. Among low-wealth households, a major factor differentiating the high-income subgroup (3 percent of total households) from their low-income counterparts (5 percent of total households) is occupation: their primary occupation is disproportionately "other than farming/ranching," whereas the low-income group was more evenly split between operators declaring farming/ranching or "other" as their primary occupation.

Do households with variable income have sufficient equity to borrow against, or to liquidate, to maintain living standards when income is low? Among farm households with income lower than the U.S. median (46 percent), household wealth exceeded the U.S. median in 9 out of 10 households. Retired households are disproportionately represented in this low-income/high-wealth subgroup. Even in the low-wealth group, basic consumption expenditures exceeded income, though to a lesser extent than for the high-wealth group.

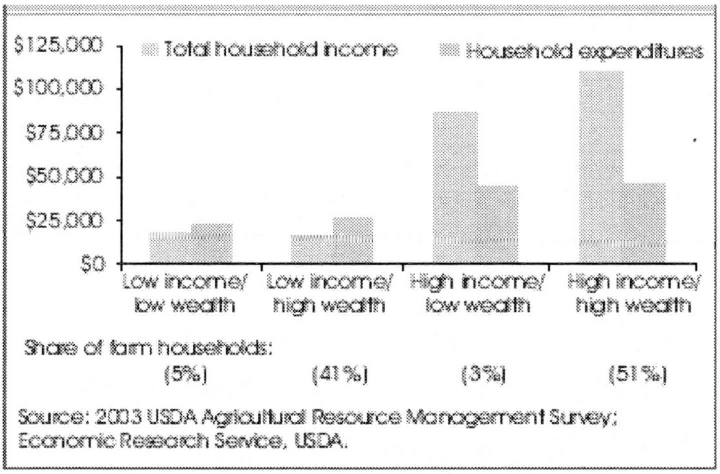

Figure 5. Average farm household income and expenditures by relative economic well-being group, 2003

Farm Households that Receive Commodity Payments also Have High Incomes and Wealth

About 32 percent of all farm households receive farm program payments, excluding environmental payments (such as those received under the Conservation Reserve Program and the Environmental Quality Incentives Program.) The share is lowest (12 percent) for households operating the smallest farms (sales less than $10,000); over 50 percent of households operating farms in each of the larger size classes receive payments. The high-income high-wealth group is more likely to be receiving program payments (34 percent) than the low-income, low-wealth group (18 percent). Among recipients, payment levels increase with production levels, and so payments disproportionately go to farm households operating larger farms, with their higher average incomes and wealth.

How We Developed the Data in this Brief

Data on farms and farm operator households are from USDA's Agricultural Resource Management Survey (ARMS). Income data for all U.S. households are from the U.S. Census Bureau's Current Population Reports, Series P-60. Wealth data for all U.S. households are derived from the Federal Reserve Board's Survey of Consumer Finances (SCF). SCF data are collected once every 3 years, and the most recent data available are from 2001. To provide a point of comparison with farm operator wealth in 2003, we estimated median wealth levels.

This brief is drawn from . . .

Ted Covey, Robert Green et al., *Agricultural Income and Finance Outlook,* AIS-83, 45 pp, November 2005.

David E. Banker and James M. MacDonald, editors, *Structural and Financial Characteristics of U.S. Farms: 2004 Family Farm Report,* Agriculture Information Bulletin No. 797, 95 pp, March 2005

James MacDonald, Robert Hoppe, and David Banker, *Growing Farm Size and the Distribution of Farm Payments*, U.S. Dept. Agr., Econ. Res. Serv., EB-6, March 2006.

End Note

[1]Operating profit refers to net farm income plus interest payments, minus the opportunity cost of operators' unpaid labor and management time.

In: America's Family Farms
Editor: Efren J. Tamayo

ISBN: 978-1-60741-751-4

Chapter 7

GROWING FARM SIZE AND THE DISTRIBUTION OF FARM PAYMENTS*

James MacDonald, Robert Hoppe and David Banker

Commodity program payments shifted sharply to higher income households between 1989 and 2003. While Congress made important changes to program design in the 1996 and 2002 farm bills, this shift was not caused by change in design. Instead, changes in payment flows resulted from structural changes in farming that are driving production to very large family farms. We expect those structural changes to continue, because larger farms appear to be more profitable and because many more operators of smaller farms are nearing retirement age. In consequence, in the absence of any fundamental changes to commodity policy, commodity payments will continue to shift to higher income households.

AGRICULTURAL PRODUCTION IS SHIFTING TO LARGER FARMS

Family-operated farms continue to account for most U.S. agricultural production. The share of production held by nonfamily farms has grown over time, but still accounted for just 14 percent of the value of production in 2003 (figure 1). A more striking shift is toward very large family farms (sales of at least $500,000, in 2003 dollars), which accounted for nearly half (45 percent) of production in 2003, up from 32 percent in 1989. The number of those very large family farms also grew—from 39,700 in 1989 to 66,700 in 2003. Meanwhile, the share of production on smaller family farms ($10,000-$250,000 in sales) fell from 40 percent in 1989 to 26 percent in 2003.

The trend to larger farms is sectorwide. The shift to larger livestock operations is well-documented and pronounced. For example, family farms with at least $500,000 in production value held 61 percent of hog production and 75 percent of poultry and egg production in

* This is an edited, reformatted and augmented version of a U. S. Department of Agriculture publication dated March 2006.

2003, up from 16 percent and 48 percent in 1989 (figure 2). But important shifts are also occurring in crop production, where very large family farms hold rapidly growing shares of production in cash grains and soybeans, tobacco, cotton, and peanuts, crops traditionally covered by commodity programs and farm legislation.

We assess changes since 1989 because we have accurate and comparable data on farm production, farm household incomes, and farm payments starting in 1989. But changes in farm structure were clearly occurring before that time; the census of agriculture shows ongoing shifts of production to larger farms in the 1970s and 1980s, continuing after the dramatic decline of farm numbers that had occurred between 1935 and 1975 had run its course. We expect these changes in farm structure to continue, for two broad reasons.

Many small farm operators are nearing retirement. Among the principal operators of smaller commercial farms, those with sales between $10,000 and $250,000, the share who are age 65 or older has risen sharply since 1989, suggesting that many are near retirement and not simply transferring the farm to younger operators. More specifically, over 30 percent of operators in the $10,000-$99,999 sales class were at least 65 years old by 2003, versus 13 percent of the operators of very large family farms.

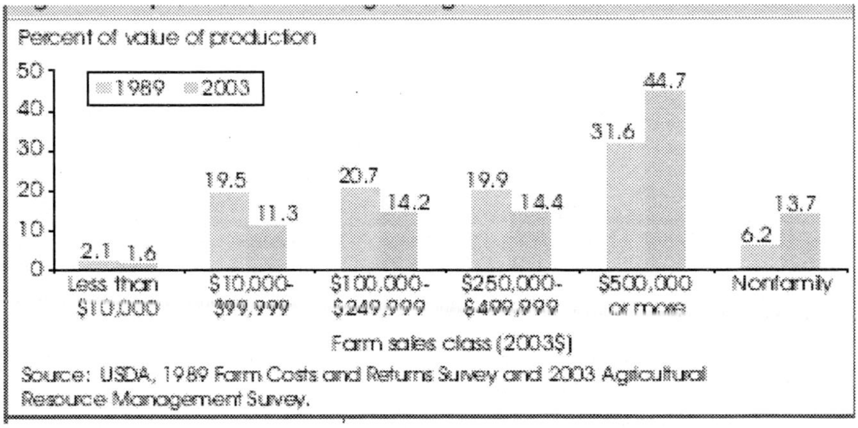

Figure 1. Agricultural production is shifting to larger farms

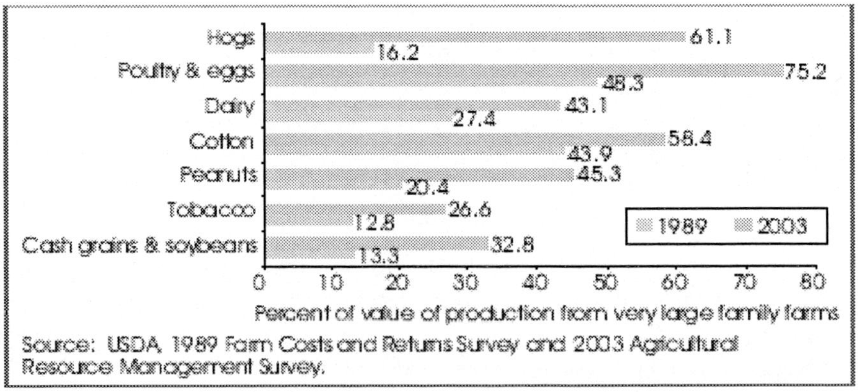

Figure 2. Crop and livestock production is shifting to very large family farms (sales of $500,000 or more in 2003$)

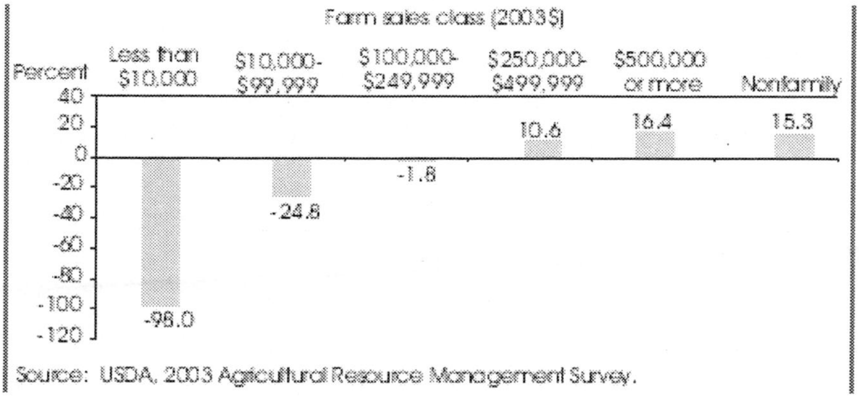

Figure 3. Operating profit margin by size of farm, 2003

Larger farms realize higher profits, on average (figure 3). Margins (the ratio of operating profit[1] to gross farm income) were negative, on average, for farms with sales below $250,000 in 2003, and they rose steadily as farm sales increased. The pattern (losses among small farms, and a strong relation between margins and farm size) holds in earlier years, and suggests that there are strong financial pressures driving production toward larger enterprises.

COMMODITY PROGRAM PAYMENTS ARE SHIFTING TO LARGER FARMS

Federal commodity programs have traditionally provided support to producers of selected commodities, principally grains and oilseeds. With production of "program commodities" shifting to larger farms, commodity payments are also shifting in that direction, since payments are linked to planting and yield histories.

Commodity payments include all commodity and disaster assistance payments, and exclude environmental payments, such as payments received under the Conservation Reserve Program (CRP) or the Environmental Quality Incentives Program (EQIP). Commodity payments reflect a farm's production history for certain commodities. Specific programs have applied to dairy, peanut, and tobacco production, while broader programs have applied to field crops such as barley, corn, cotton, oats, rice, sorghum, soybeans, and wheat. Payments are tied to the amount of a farm's cropland that has been enrolled in programs, as well as yield histories. As a result, farms that produce higher volumes of program commodities generally receive higher payments.

High-value crops, as well as fed cattle, hogs, and poultry, are not supported by traditional government price and income support programs, although they do receive disaster assistance and occasionally may benefit from an ad hoc support program. Consequently, farms that produce such commodities receive substantial commodity payments only if they also produce program commodities or have a history of producing them. Because production of fed cattle, hogs, poultry, and high-value crops tends to occur on very large farms, farms in that sales category have traditionally drawn relatively small shares of total commodity payments.

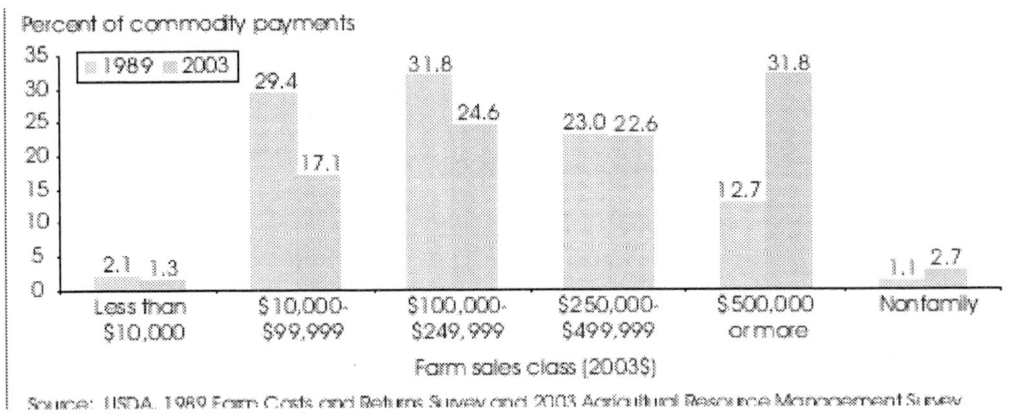

Figure 4. Government commodity payments are shifting to larger farms

However, as production of traditional program commodity crops shifted to very large farms, commodity payments also shifted sharply (figure 4). Farms with less than $250,000 in production value (2003$) received 63 percent of commodity payments in 1989; by 2003, they received 43 percent of payments. But farms with at least $500,000 of production received 32 percent of all commodity payments in 2003, up from 13 percent in 1989.

ABOUT THE DATA

Data on farms and farm operator households were obtained from USDA's Agricultural Resource Management Survey (ARMS) and its predecessor, the Farms Costs and Returns Survey (FCRS). The FCRS first reported comparable data in 1989, so we used that year as our starting point. Our classification sorts family farms among five sales classes, and assigns all nonfamily farms to a sixth class. Income data for all U.S. households were obtained from the U.S. Census Bureau's Current Population Reports, Series P-60. We accounted for changes in farm product prices by converting all farm sales figures to 2003 dollars, using the Producer Price Index for Farm Products, and we converted all household income figures to 2003 dollars using the Consumer Price Index (CPI-U), in order to account for changes in the purchasing power of incomes.

OPERATORS OF THE LARGEST FARMS HAVE HIGHER INCOMES

Farm households are not, in general, poor. Mean household income among all U.S. farm operator households was $68,500 in 2003, which compares favorably to the nationwide mean household income of $59,100 (for more on how household incomes are distributed, see Economic Brief No. 7, *Economic Well-Being of Farm Households*). Moreover, mean incomes do not vary sharply with farm size among smaller farms, those with sales below $250,000. Operators of the smallest farms, those with less than $10,000 in sales, derive almost all of their household income from off-farm work and from unearned income, such as Social Security, pensions, and financial investments. About 75 percent of those operators report negative incomes from farming, but those losses are offset by enough off-farm income to raise their household incomes above the national average. Households that operate farms with annual sales up to $250,000 frequently combine a financially viable farm business with offfarm employment to generate household incomes that match or exceed national averages.

By 2003, half of commodity payments went to households with income above $75,772.

However, operators of very large family farms realize much higher incomes, on average, and as production shifts to those farms, it also shifts to much higher-income farm households. The principal operators of very large family farms reported a mean household income of $214,200 in 2003, well above the mean across all family farms, and well above the mean of $102,400 reported by principal operators in the next largest size class ($250,000-$500,000).

COMMODITY PAYMENTS ARE SHIFTING TO HIGHER-INCOME HOUSEHOLDS

In 1989, half of commodity payments went to principal operators whose households earned more than $45,808 (in 2003 dollars), and half went to principal operators whose households had incomes below that figure (table 1). To further summarize the distribution of payments in 1989, one quarter went to households earning more than $94,784 (also in 2003 dollars), while 10 percent went to households with incomes above $189,149.

Since then, payments have shifted sharply to higher-income farm households. By 2003, half of commodity payments went to households with income above $75,772. One quarter went to households earning more than $160,142, and 10 percent of payments went to households earning more than $342,918.

Table 1. Commodity Payments Are Shifting to Higher-Income Households

	1989	1991	1997	2003	1989-2003 change
		Household Income (2003$)			*Percent*
Commodity Payments Distribution					
50th percentile (median)	45,808	47,121	55,607	75,772	65.4
75th percentile	94,784	98,657	122,868	160,142	69.0
90th percentile	189,149	196,442	250,092	342,918	81.3
All U.S. Households					
50th percentile (median)	42,892	40,686	42,425	43,318	1.0
90th percentile	107,580	103,394	112,589	118,200	9.9

Note: Half of commodity payments go to households with incomes higher than the 50th percentile; a quarter go to households with incomes higher than the 75th percentile, and a tenth go to households with income higher than the 90th percentile.

Because household incomes did not rise sharply with farm sales among operators of farms with less than $500,000 in sales, shifts in production to larger farms within these size classes did not shift commodity payments to noticeably higher income households. Rather, the apparent shift in commodity payments to higher income households is being driven by shifts of production to the largest class of farms (over $500,000 in sales), whose households have substantially higher incomes.

These shifts have far outpaced the growth in overall U.S. incomes. Between 1989 and 2003, median U.S. household income grew by just 1 percent, from $42,892 (2003 dollars) to $43,318. The median U.S. household income in 1989 was near the median of the farm payments distribution ($45,808). This was not so by 2003.

In the last two decades, incomes have grown most rapidly at the upper levels of the income distribution; at the 90th percentile (10 percent of households earn more), U.S. income grew by 10 percent between 1989 and 2003. The incomes of households receiving most commodity payments have grown even more sharply than this. In short, commodity payments are being shifted, through structural change, toward relatively high-income households.

See also . .

Farm Structure briefing room, ERS website, www.ers.usda.gov/Briefing/ FarmStructure/

Carol A. Jones, Hisham El- Osta, and Robert Green. *Economic Well-Being of Farm Households*, U.S. Dept. Agr., Econ. Res. Serv., EB-7, March 2006.

End Note

[1]Net farm income plus interest payments, minus the opportunity cost of operators' unpaid labor and management time.

In: America's Family Farms
Editor: Efren J. Tamayo

ISBN: 978-1-60741-751-4

Chapter 8

OFF-FARM INCOME, TECHNOLOGY ADOPTION, AND FARM ECONOMIC PERFORMANCE*

Jorge Fernandez-Cornejo, Ashok Mishra, Richard Nehring, Chad Hendricks, Malaya Southern and Alexandra Gregory

ABSTRACT

The economic well-being of most U.S. farm households depends on income from both onfarm and off-farm activities. Consequently, for many farm households, economic decisions (including technology adoption and other production decisions) are likely to be shaped by the allocation of managerial time among such activities. While time allocation decisions are usually not measured directly, we observe the outcomes of such decisions, such as onfarm and off-farm income. This report finds that a farm operator's off-farm employment and off-farm income vary inversely with the size of the farm. Operators of smaller farm operations improve their economic performance by compensating for the scale disadvantages of their farm business with more off-farm involvement. Off-farm work reduces farm-level technical efficiency, but increases household-level technical efficiency. And adoption of agricultural innovations that save managerial time is associated with higher off-farm income.

Keywords: Off-farm income, farm households, economic performance, managerial time, scale economies, scope economies, technical efficiency, technology adoption, farm size.

ACKNOWLEDGMENTS

The authors thank James MacDonald, Keith Wiebe, Dayton Lambert, Carol Jones, and Utpal Vasavada, ERS, for the helpful comments provided on earlier drafts of the report. We

* This is an edited, reformatted and augmented version of a U. S. Department of Agriculture publication dated January 2007.

also thank James Hrubovcak from the Office of the Chief Economist, Jane Schuchardt from the Cooperative State Research, Education and Extension Service, Bruce Gardner from the University of Maryland, and an anonymous reviewer. Finally, we are very grateful to Dale Simms for his valuable and prompt editorial assistance and Anne Pearl for cover design and document layout.

SUMMARY

U.S. farmers must make a host of decisions relating to their farms' operation, including what to grow, when to grow it, in what quantities, and by what methods. Often overlooked in this calculation, but factoring heavily in the diversity of U.S. farms and farm households, is the fact that most operators split their time between farm and nonfarm activities. Large farms are typically able to economize on inputs and better coordinate stages of production. Smaller farms, though often unprofitable from a farm business perspective, have endured by being part of household enterprises that combine farm and off-farm activities. Their operators' onfarm decisions, from choice of technology to choice of specialty, are often influenced by off-farm commitments and income.

What Is the Issue?

Onfarm and off-farm activities compete for limited managerial time (mainly of the operator and spouse). How farm operator households allocate their time largely affects production decisions (such as technology adoption), economic performance, and the household's economic well-being.

The extent of off-farm work and its relationship with farm economic performance may have important policy implications. For example, government policies for agriculture (via conservation, research and development, extension, and commodity programs) may affect farm households differently, depending on the relative importance of onfarm versus off-farm income. And the effectiveness of policies promoting adoption of farm technologies might be improved by taking into account the different demands on managerial time and the relative ability of the farm household to accommodate those demands.

What Did the Study Find?

Operators of smaller farms typically participate more in off-farm employment, work more hours off the farm, and have higher off-farm income than operators of larger farms. In 2004, farm households with farm sales less than $10,000 had average off-farm earned income of $54,600, while households with farm sales of $500,000 – $1 million averaged only $30,100. More than 58 percent of operators with farm sales less than $10,000 reported off-farm hours worked in 2004, versus less than 20 percent for operators of farms with sales of $500,000-$ 1 million.

As previous studies have shown, **off-farm work is less likely on farms with labor-intensive enterprises** such as dairy. Moreover, dairy farmers who do work off the farm tend to require higher compensation to do so than farmers producing other commodities. Off-farm work has also been shown to be positively related to urban proximity and to the education and experience of the operator and spouse.

Including off-farm income-generating activities improves the overall economic performance of the farm household. Off-farm income clearly adds to total household income, but it can also improve efficiency and other measures of performance of the farm household. Our estimates for corn and soybean farms show that households engaged in off-farm income-generating activities together with the production of traditional farm outputs have cost savings of 24 percent relative to carrying out those activities separately. The savings likely arise from the sharing of managerial expertise (and its many components, such as accounting and information processing skills, sales expertise, administrative and technical know-how, etc.) between onfarm and off-farm activities. For example, management skills acquired in farming might be applicable to (and shared with) a nonfarm business, and vice-versa.

From a farm business perspective, operators of smaller farms have a greater **incentive to expand**. However, from a household perspective (including off- farm income-generating activities), operators of small farms have a reduced tendency to increase their farm size.

Large farms are generally more efficient than smaller farms in transforming farm inputs into outputs, given the technology at their disposal. But focusing on farm inputs and outputs alone is misleading because off- farm income-generating activities are increasingly important in determining economic performance of the farm household.

When off-farm activities are included, farm household-level efficiencies are higher than farm-level efficiencies across all farm sizes, and efficiency gains from integrating off-farm work into the output portfolio are relatively greatest for smaller farms. As a result, household-level efficiencies of smaller farms are comparable to farm-level efficiencies of larger farms.

This suggests that households operating small farms have partially adapted to shortfalls in farm-level performance by increasing their off-farm income.

In addition to its links with the farm business, as traditionally examined, farmers' technology choices are closely related to off-farm income. Higher off-farm income is significantly related to the adoption of technologies that economize on management time (management saving such as herbicide-tolerant crops, conservation tillage). For example, a 16-percent increase in off-farm household income is associated with a 10-percent increase in the probability of adopting herbicide-tolerant (HT) soybeans. Household income from onfarm sources is not significantly associated with adoption of these technologies, but total household income (including income from off-farm sources) is. On the other hand, lower off-farm income is significantly related to adoption of managerially intensive technologies (such as precision farming). For example, an 8-percent decrease in off-farm income is associated with a 10-percent increase in the probability of adopting yield monitors, a key component of precision agriculture. These findings corroborate a tradeoff between household/operator time spent in onfarm and off-farm activities. Households operating small farms devote more time to off-farm opportunities and are more likely to adopt management-saving technologies.

How Was the Study Conducted?

To examine the relationships between off-farm income, farm and household characteristics, and economic performance of U.S. farm households, we developed econometric models and estimated them using USDA's Agricultural Resource Management Survey (ARMS) data for several years (1996- 2001). To examine the relationship between off-farm work and economic performance of farm households (including economies of scale and scope, and economic efficiency), we compared estimates obtained using traditional farm-level models to estimates obtained using household-level models (including off-farm income-generating activities along with traditional farm outputs such as crops and livestock). To examine the relationship between off-farm income and technology adoption, we developed a model that incorporates the adoption decision into the agricultural household framework. We examined the interaction of off-farm work and adoption of agricultural technologies of varying managerial intensity, including herbicide-tolerant crops, precision agriculture, conservation tillage, and Bt (*Bacillus thuringiensis)* corn, after controlling for other factors.

INTRODUCTION AND OVERVIEW

Decisionmakers (mainly farm operators and their spouses) are a major determinant of farms' economic performance. The effort and ability to manage land, water, machinery, and other inputs— as well as adoption of technologies and production practices—can help secure farm business success and the economic well-being of a farm household. However, many farm operators (and other household members) use a large share of their time in off-farm income-generating activities. Consequently, for many farm households, economic decisions (including technology adoption and other production decisions) are likely to shape and be shaped by the allocation of managerial time to such activities. While time allocation decisions are usually not measured directly, we observe the outcomes of such decisions, such as onfarm and off-farm income.

Off-farm income (largely earned income from employment and off-farm business income) received by U.S. farm operators and their spouses has risen steadily over recent decades and now constitutes the largest component of farm household income (figure 1a, b). The impact of off-farm income is felt particularly by households operating small farms, allowing many of them to survive and even flourish to an extent not thought possible 20 or 30 years ago (Gardner, 2005). In addition, the growth in off-farm income over the last 40 years reduced income inequality among farm households and helped U.S. farmers' average incomes overtake those of the nonfarm population (Gardner, 2002).

This report examines the empirical relationships between off-farm income, farm household characteristics, production decisions (particularly technology adoption), and various measures of economic performance for U.S. farm households. This research provides insights into farmers' choices in the context of farm/household integration and helps improve our understanding of the pace of technological innovation and its relation to the structure of agriculture.

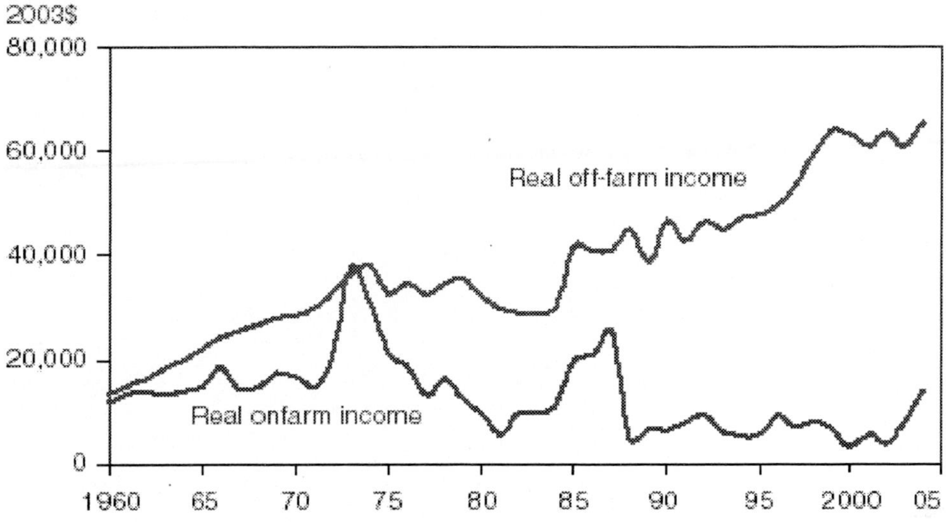

Sources: USDA, ERS. Deflator used to calculate real income is the consumer price index (CPI-U) from the Bureau of Labor Statistics.

Figure 1a. Farm household income, U.S. average 1960-2004

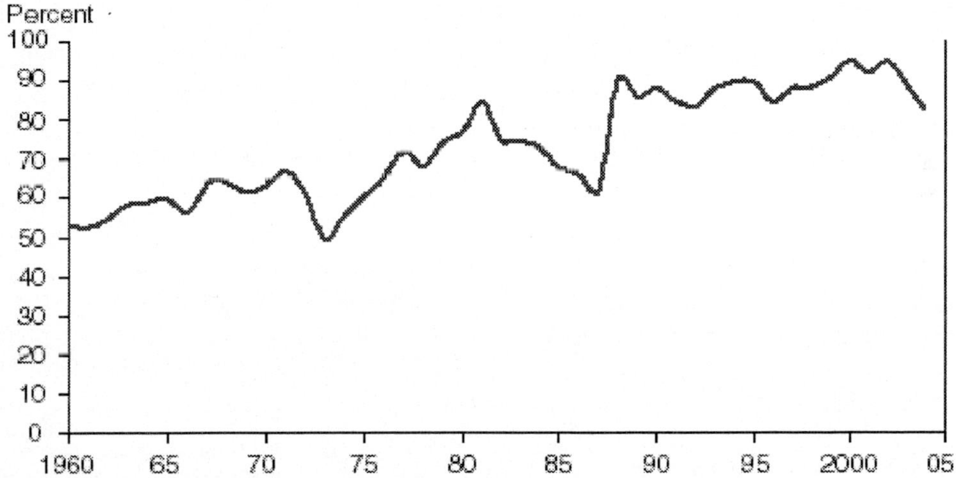

Sources: 1960-2003: USDA farm household income estimates over time, ERS farm structure briefing room, http://www.ers.usda.gov/Briefing/FarmStructure/Data/historic.htm; 2004: Covey et al., 2005.

Figure 1b. Off-farm income share of total farm household income, U.S. average, 1960-2004

The report also suggests the need to analyze the economics of the farm business and farm household in an integrated framework and describes two approaches for doing so. We summarize statistics of off-farm work and income in U.S. farm households and examine the relationship between off- farm income and farm size, location, and household characteristics.

Our main research focus is to examine how off-farm work influences the economic performance of the integrated farm business and household. To do this, we expand traditional concepts of economic performance, such as economies of scale and efficiency, to incorporate onfarm and off-farm income-generating activities of household members. In addition, we examine the relationship between off-farm income and the adoption of agricultural technologies of varying managerial intensity, namely herbicide-tolerant crops, precision agriculture, conservation tillage, and Bt (*Bacillus thuringiensis*) corn.

An Integrated Approach

While increasing household income, off-farm activities also compete for managerial time (mainly of farm operators and their spouses), which may affect the economic performance of the farm business. Consequently, economic decisions (including technology adoption and other production decisions) are likely to shape and be shaped by the underlying allocation of time within the farm operator household. So, rather than examining the farm business or farm household in isolation, an integrated approach captures the interplay of farm and nonfarm considerations and contributions.

Despite its importance, the role of off-farm income has been largely neglected in empirical analyses of farm economic performance and technology adoption.[1] Some exceptions include Gardner (2001), Boisvert (2002), Goodwin and Mishra (2004), Fernandez-Cornejo et al. (2005), Nehring et al. (2005), Paul and Nehring (2005), and Chavas et al. (2005).[2] One reason for this lack of studies may be the modeling and data challenges in moving from the traditional unit of analysis (the farm business) to the farm household.

While agricultural economists have made major contributions in understanding farm production functions, they may not have exploited as fully the concept of the household production function (Offutt, 2002). In this context, the allocation of time (and money) of household members to production, consumption, and other activities is particularly important. An integrated firm-household perspective was suggested back in 1952 by E.O. Heady, who observed that "the firm-household complex is important not only to defining the organization of resources and family activities which will maximize utility at a given point in time but also in helping to explain uncertainty precautions, capital accumulation, soil conservation, and other production-consumption decisions, which relate to time."[3]

Approaches to Integrate Off-Farm Work and Farm Production

Two approaches are used in this report to model the interaction of off-farm income-generating activities with traditional farm production activities. The unifying notion underlying the two approaches is that managerial time is a key resource in both onfarm and off-farm activities.

In one approach, we expand the agricultural household model to include the technology adoption decision together with the off-farm work decisions by the operator and spouse. The agricultural household model describes how a farm household allocates its time (and other resources) among producing commodities, earning off-farm income, leisure, and home production.[4] The model assumes that the farm household maximizes its utility subject to constraints on its time (including work and leisure), income, and production technology (production function). Household members derive utility from goods purchased for consumption, leisure, and factors exogenous to current household decisions, such as human capital, household characteristics, and weather. Using this model, we examine the interaction of off-farm work and the adoption of agricultural innovations (both management saving like herbicide- tolerant crops, and management using like precision agriculture or integrated pest management—IPM), then obtain empirical estimates of the relationship between adoption of these technologies and farm household income.

Though the agricultural household model has intuitive appeal in modeling farm household behavior, it requires much in the way of assumptions and data (Offutt, 2002). Parameter estimation for the models spawned by the household production function often requires hard-to-get data, including consumption expenditures, farm and off-farm labor supply, farm and nonfarm outputs and inputs, assets, and prices for all goods, inputs, and labor. Also needed is information on technologies and participation in government programs, as well as demographic data. For these reasons, it is sometimes necessary to use alternative methods. In this approach, we expand the concept of scope economies to include as output all income- generating activities, on or off the farm, in addition to the traditional farm outputs such as corn, soybeans, and livestock (Nehring et al., 2005). In addition, we estimate scale economies and technical efficiency, and compare results at the farm and household levels.

SCALE AND EFFICIENCY

Scale Economies

A farm is said to have economies of scale (or increasing returns to scale) if the average cost declines as output (scale of production) increases. If a farm is subject to economies of scale, it is cost effective for that farm to increase all outputs simultaneously while holding the mix of outputs constant (costs would rise less than proportionally). Thus, the existence of scale economies suggests that farms can achieve lower average costs by becoming larger. Economists have established (under reasonable conditions) the equivalence between the information provided by the costs and the production technology (Carlton and Perloff, 2000). Based on the production technology, economies of scale may be viewed from an output or input perspective.

From an output perspective, the term elasticity of scale is used to measure the percent increase in output generated by a 1-percent increase in all inputs (Varian, 1992). There are increasing returns to scale if the elasticity is greater than 1; that is, an increase in overall inputs generates a more than proportionate increase in output. For example, a scale elasticity of 1.15 means that a 1-percent increase in inputs leads to a 1.15- percent increase in output. Conversely, if the elasticity is lower than one there are decreasing returns to scale; that is, an increase in overall inputs generates a less than proportionate increase in output. For example, a scale elasticity of 0.8 means that a 1-percent increase in inputs leads to a 0.8-percent increase in output. Constant returns to scale means that a 1-percent increase in overall inputs generates a 1-percent increase in output; in this case the elasticity of scale is equal to 1.

From an input perspective, a similarly defined scale elasticity measures the percent increase in inputs required to support a 1-percent increase in all outputs. In this case, returns to scale are increasing when the input-oriented scale elasticity is less than one. For example, if the scale elasticity of a farm is 0.75, it means that a 0.75-percent increase in inputs will be needed to support an output increase of 1 percent. This suggests that there is an incentive for the farm to grow larger. If the elasticity is equal to one (constant returns to scale), there are no scale economies available. In this report, we use an input perspective (input distance function, appendix 1).

Technical Efficiency

Economic efficiency can be decomposed into technical efficiency and allocative efficiency. A farm is technically efficient if it uses the minimum possible levels of inputs to produce a given level of output, given the technology. An allocative efficient farm produces a given output using the best (minimum cost) input proportions given prevailing input prices. Unless specified otherwise, the efficiency results discussed in this report involve technical efficiency.

Technical efficiency is the ratio of current to maximum possible or "best practice" production and it is calculated in this study using an input distance function (see appendix 1). Technical efficiency is defined relative to an "efficient frontier" and all farms operating on the efficient frontier are classified as 100 percent efficient with an efficiency score equal to 1. Farms using more inputs to produce a given output level than those on the efficient frontier are inefficient and their efficiency score is less than 1. Technical efficiency is often associated with managerial ability and experience.

OFF-FARM WORK AND INCOME IN U.S. FARM HOUSEHOLDS

Off-farm income received by farm operators and their spouses has risen steadily over recent decades (figure 1a) as job opportunities have grown and technological progress, such as mechanization, has lessened onfarm labor needs. The off-farm income share of total household income of U.S. farmers rose from about 50 percent in 1960 to more than 80 percent over the past 10 years (figure 1b). Most of the off-farm income was earned. On average, a farm household earned about $48,800 from off-farm sources in 2004, received about $18,500 in unearned income (Social Security, interest, etc), and netted nearly $14,200 from farming activities (Covey et al., 2005).[5] Fifty-two percent of farm operators worked off farm in 2004 (up from 44 percent in 1979). The share of spouses working off farm grew from 28 percent of spouses in 1979 to 45 percent in 2004 (Mishra et al., 2002; 2004 ARMS data).

The trend is similar in terms of hours worked (table 1). Average hours worked off farm by farm operators has increased (from 830 hours per year in 1996 to 1,022 in 2004), while the hours devoted to farm work did not change markedly (1,525 hours in 1996 and 1,574 in 2004). Similarly, the number of hours worked off the farm by spouses increased from 690 in 1996 to 809 in 2004.

Farmers' Motivations to Work Off Farm

Once seen as a "temporary response to the Great Depression," off-farm employment is now regarded as a "regular feature of almost all farming societies" (Fuller, 1991; Bartlett, 1986; Bessant, 2000). More than half of U.S. farm operators now work off the farm.[6] Moreover, off-farm income appears to smooth out household income flows (Mishra and Goodwin, 1997; Mishra and Sandretto, 2002), and most farmers view off-farm employment as a permanent rather than a temporary or transitional (into or out of farming) pursuit (Ahearn and El-Osta, 1993).[7] Farm operators in a 1982 survey felt that full-time farming provided inadequate income (91 percent of the respondents), and that farm income was risky (70 percent) and offered no fringe benefits such as pensions and health insurance (55 percent). Capital and land constraints were considered less important disadvantages to full-time farming (42 and 30 percent) (Barlett, 1991). More recently, the 2004 ARMS asked operators and spouses to list the two main reasons for seeking off-farm work. The primary reason given by 35-50 percent of the operators and 44-63 percent of the spouses (depending on farm size and occupation of the farm operator) was "to increase income" of the farm household. Other reasons cited were to obtain fringe benefits (such as health insurance) and personal satisfaction (Covey et al., 2005).

So most operators and spouses report working off farm primarily to increase income for the farm household, but how was the additional income used? Contrary to conventional wisdom, most farm operators and spouses did not work off the farm to directly support their farm business. USDA surveys indicate reasons unrelated to the farm business, from buying groceries to funding a retirement account (Hoppe, 2001).

Table 1. Operator and spouse hours worked on and off farm, 1996-2004

Item	1996	2000	2004
Operator hours worked:			
On farm	1,525	1,433	1,547
Off farm	830	1,011	1,022
Total	**2,355**	**2,443**	**2,596**
Spouse hours worked:			
On farm	366	337	877
Off farm	690	751	809
Total	**1,056**	**1,089**	**1,686**

Sources: 1996: Hoppe (2001, p. 29); 2000: Mishra et al. (2002, p. 50); 2004: ARMS data.

Farmers and spouses hold a variety of off-farm jobs, but especially in private businesses (54.1 percent of operators with off-farm jobs), self-employment (22.3 percent), and government (16.0 percent). Only 3.3 percent worked on another farm (Mishra et al., 2002).

Spouses with off-farm work are most likely to be employed in the private sector (55.1 percent) and government (28.4 percent), with less than 1 percent working on another farm.

Opportunity Cost of Labor for Farm Operators

Opportunity cost is an important economic concept that measures the economic cost of an action or decision in terms of what is given up (opportunity forgone) to carry out that action. In the case of farm labor, for example, the opportunity cost of labor for the operator (or spouse) labor is often measured in terms of the wage that the operator (or spouse) can obtain working off farm. As the United Nations' Economic Commission for Europe notes: "In conventional accounting systems, 'unpaid' family labour does not usually appear as an explicit cost of production. Consequently, there is no explicit 'wage' paid to the labour that the farmer and his family [contribute to] production."

Farm household labor is a critical input in agricultural production. In the corn/soybean-producing States, farm household members provide more than 80 percent of all labor hours.[8] A significant proportion of those labor hours is not valued directly in the marketplace (e.g., through wages). Studies have estimated the opportunity costs of farm labor by using predicted off-farm wages (El-Osta and Ahearn, 1996).

Alternatively, a simplified approximation of the opportunity cost of labor for farm operators and their spouses can be obtained directly from ARMS data. The (nominal) opportunity costs for corn/soybean operators and spouses appear not to have increased over 1996-2000. The cost for the operator ($21.07 per hour for 2000) appears to run about 20 percent higher than that of the spouse, and both are higher than the actual wage rate for hired farm labor.[9]

It is also interesting to compare the opportunity cost of labor for corn/soybean farmers with those of dairy farmers. The cost for U.S. dairy farmers in 2000 was econometrically estimated at $27.58 per hour for operators (30 percent higher than for corn/soybean farmers) and $19.36 for spouses (18 percent higher) (Lovell and Mosheim, 2005). Given that labor requirements in dairy production are high and inflexible (El-Osta and Ahearn), dairy farmers likely require a higher "wage" to work off the farm than farmers working in other enterprises.

Table 2. Opportunity cost of labor for corn/soybean farm operators and spouses, and actual hired farm wage rate, 1996-2000

Year	Operator	Spouse	Hired
	Dollars per hour		
1996	22.88	17.87	7.42
1997	26.72	19.06	8.01
1998	22.14	18.77	8.30
1999	22.19	17.96	8.67
2000	21.07	17.47	8.99

Source: ERS estimates based on ARMS data for corn/soybean States analyzed (Nehring, Fernandez-Cornejo, and Banker, 2005).

OFF-FARM INCOME AND FARM/HOUSEHOLD CHARACTERISTICS

Like their nonfarm counterparts, many farm households are dual career. While operators and spouses across all sizes and typologies work off-farm or manage nonfarm businesses, the level of off-farm income varies with farm size, region, farm type, and the human capital of operators and spouses.

Off-Farm Income and Farm Size

Off-farm income varies inversely with farm size; operators of smaller farms have higher off-farm incomes, both earned and total.[10] Farm households with gross farm sales less than $10,000 had total off-farm income averaging nearly $74,000 in 2004 ($54,600 of which was earned), while households with farm sales between $250,000 and $499,999 had total off-farm income averaging about $45,000 ($33,200 earned) (table 3). While off-farm income constitutes the largest component of farm household income on average, its share decreases with farm size. For farms with gross sales higher than $250,000 (less than 8 percent of U.S. farms), off-farm income is no longer the largest component of household income (table 4).

Off-farm household income earned by the operators is more variable across farm sizes ($27,500 for operators of smaller farms versus less than $10,000 for operators of the largest farms) than that earned by spouses (between $12,000 and $14,000 across all sizes in 2004). Off-farm income earned by other household members averages around $1,000.

To a large extent, the inverse relationship between off-farm earned income and farm size is due to greater off-farm employment (and more hours worked off the farm) by operators of smaller farms. More than 55 percent of operators with farm sales less than $100,000 reported off-farm hours in 2004 versus 20 percent or less for operators of farms with sales above $250,000 (table 4). On the other hand, off-farm income earned by farm operators who work off-farm does not vary much with size, averaging $47,000 for operators of the smallest farms and $39,000 for operators of the largest farms.

Table 3. Off-farm household income by farm size, 2004

Farm sales	Share of farms	Income earned by the operator	Income earned by the spouse	Income earned by other members	Off-farm business income	Total earned income	Unearned income	Total off-farm income
	Percent		*Dollars*					
$9,999 or less	43.7	27,457	14,756	1,219	11,209	54,641	19,392	74,033
$10,000-$99,999	40.7	24,295	13,095	1,142	9,889	48,422	19,549	67,971
$100,000-$249,999	7.9	11,074	14,722	1,158	8,493	35,445	11,467	46,913
$250,000-$499,999	4.2	7,559	13,439	836	11,404	33,238	11,633	44,870
$500,000-$999,999	2.0	7,790	12,816	1,110	8,371	30,086	21,991	52,077
$1,000,000 or more	1.5	4,898	12,017	612	10,744	28,271	12,811	41,082
All farms	100.0	23,318	13,943	1,156	10,402	48,818	18,461	67,279

Source: 2004 ARMS data.

Table 4. Farm household income by farm size, 2004

Farm size (annual sales)	Number of farms	Share of farms income	Total household farming	Income from income	Share of farm	Off-farm income	
	Number	*Percent*	*Dollars*	*Dollars*	*Percent*	*Dollars*	
$9,999 or less	901,333	43.7	71,155	-2,878	-8.9	74,033	
$10,000-$99,999	838,912	40.7	72,061	4,091	11.7	67,971	
$100,000-$249,999	162,782	7.9	80,912	33,999	18.9	46,913	
$250,000-$499,999	86,087	4.2	124,386	79,516	23.4	44,870	
$500,000-$999,999	41,424	0.2.0	168,844	116,766	16.5	52,077	
$1,000,000 or more	30,284	1.5	411,266	370,184	38.3	41,082	
All farms	2,060,822	100.0	81,480	14,201	100.0	67,279	
Farm size (annual sales)	Earned off-farm income	Share of operators reporting off-farm hours	Off-farm earned income by operators who worked	Off-farm earned income of operators	Share of spouses reporting off-farm hours	Off-farm income earned by spouses	Off-farm earned income of spouses who worked off-farm
	Dollars	*Percent*	*Dollars*	*Dollars*	*Percent*	*Dollars*	*Dollars*
$9,999 or less	54,641	58.7	27,457	46,775	44.1	14,756	33,460
$10,000-$99,999	48,422	55.5	24,295	43,775	45.5	13,095	28,780
$100,000-$249,999	35,445	31.1	11,074	35,608	54.4	14,722	27,063
$250,000-$499,999	33,238	20.4	7,559	37,054	45.2	13,439	29,732
$500,000-$999,999	30,086	18.6	7,790	41,882	44.8	12,816	28,607
$1,000,000 or more	28,271	12.6	4,898	38,873	37.2	12,017	32,304
All farms	48,818	52.1	23,318	44,756	45.4	13,943	30,711

Source: 2004 ARMS data.

The inverse relationship between farm size and off-farm work still holds after controlling for other factors, as demonstrated econometrically by many researchers (Lass et al., 1989, 1991; Yee et al., 2004). In addition, Goodwin and Bruer (2003) and Fernandez-Cornejo et al. (2005) showed that the inverse relationship holds for both operator and spouse.

Time allocation between onfarm and off-farm activities by household members appears to be the underlying reason for the inverse relationship between farm size and off-farm work. This relationship appears to be valid regardless of the sequence in which time is allocated between farm and off- farm work. As Olfert (1984) notes, it may be the case that farmers choose farm size and type after knowing the time commitments required by an off- farm job, or farmers may choose the type and amount of off-farm work after taking into account the nature of the labor requirements on the farm.[11]

Off-Farm Income and Farm Location

Off-farm employment also varies geographically, with widely differing shares of off-farm income (to total income) even within States (figure 2). In general, high ratios of off-farm earned income to total income are exhibited in the four regions—the Northeast, Appalachian, Southern Plains, and Northwest—where job opportunities tend to be highest or farm income lowest. In many cases, one family member may focus on the farm operation while the spouse and children work off the farm. In other situations, the farm operation may be a side job and a refuge from urban stress.

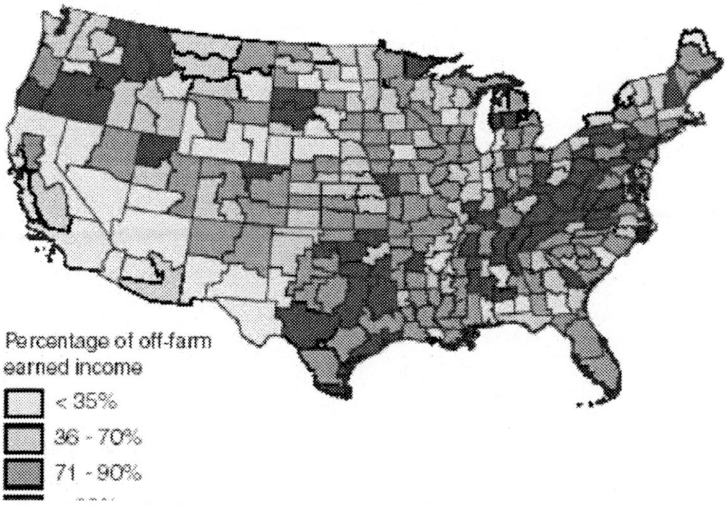

*ASD = Agricultural Statistics District. Source: 2001 ARMS data.
Source: 2001 ARMS data.

Figure 2. The importance of off-farm income by ASD*, 2001 (off-farm earned income/total income)

The supply of off-farm labor has been shown to be positively related to urban proximity (Lass et al., 1991). Moreover, Gardner (2001) found that farmers' income growth is inversely related to the rural share of a State's population. Gardner observed that this finding supports Schultz's (1950) hypothesis that "a larger presence of nonfarm people in a State is good for the growth of farmers' incomes, because it increases their off-farm earnings opportunities and increases the demand for the goods and services that farmers produce." This may be particularly true for agricultural States with large urban populations such as Texas, where off-farm opportunities increase near one of that State's four major cities—Dallas-Fort Worth, Houston, San Antonio, and Austin.

Off-Farm Income, Type of Enterprise, and Human Capital

Off-farm work is less likely on farms with labor-intensive enterprises such as dairy (Leistritz et al., 1985) and other livestock (Lass et al., 1991; Goodwin and Bruer, 2003). Moreover, dairy farmers who do work off the farm tend to require higher "wages" (the opportunity cost of labor is higher) to work off farm than farmers working in other enterprises.

The supply of off-farm labor has also been shown to be positively related to human capital such as education and experience of the operator and spouse (Lass et al., 1991). The number of children is positively associated with off- farm employment for farm men, but the association is negative for farm women. More children may imply more need for additional income but also additional child care at home.

OFF-FARM WORK, SCALE AND SCOPE ECONOMIES, AND EFFICIENCY

The importance of off-farm income to all U.S. farmers is widely acknowledged, and the relative dedication to off-farm work is related to farm size, location, specialty, and operator characteristics. However, is off-farm work actually helping farm households in general, and those operating small farms in particular, to improve their economic performance? Since scale and scope economies, as well as economic efficiency, are key concepts used by economists to examine economic performance, this section introduces those concepts as they relate to off-farm work.

A farm is said to have economies of scale (or increasing returns to scale) if the average cost of production declines as output (scale of production) increases (see box, p. 4). This decline in per-unit costs as output increases suggests that smaller farms can achieve cost advantages by becoming larger. The concept of economies of scale is an important one. For example, farms with lower average costs are better able to cope with higher input prices (Kumbhakar, 1993). On the other hand, increasing returns to scale in production may lead to consolidation of firms with potential effects on competition (Hallam, 1991).

With multiple outputs, the measurement of scale economies becomes more complicated. In addition to changes in costs that occur as output expands, there are also changes in costs due to the product mix (Hallam, 1991). If it is cheaper to produce several outputs in one operation than it is to produce them in separate operations, economies of scope are said to occur (see box, p. 14).

Table 5. Scale economies for corn/soybean farms, 1996-2000

		Elasticity of scale	
Farm type[1]	Gross sales	Farm level (Excluding off-farm income)	Household level (Including off-farm income)
Farming occupation/ lower sales	< $100,000	0.56	0.73
Farming occupation/ medium sales	$100,000-$249,999	0.74	0.88
Large family farms	$250,000-$499,999	0.77	0.94
Vary large family farms	>$500,000	0.80	0.96
All farms		0.66	0.83

[1]Excluding limited-resource farms and retirement/residential farms. Limited-resource farms are small farms with gross sales less than $100,000, total farm assets less than $150,000, and total operator household income less than $20,000. Limited-resource farmers may report farming, a nonfarm occupation, or retirement as their major occupation. Retirement/residential farms are small farms whose operators report they are retired or engaged in a major occupation other than farming) (Hoppe et al., 1999)

Source: Nehring et al., 2005.

Off-Farm Work and Scale Economies

We estimated the scale economies for corn and soybean farms for 1996-2000, from an input perspective. Scale economies both at the farm level (the measure traditionally reported) and at the household level (including off-farm income-generating activities as an output) are considered. At the farm level, the elasticity of scale ranges from about 0.56 for smaller farms (gross sales less than $100,000), to about 0.8 for the larger farms (sales greater than $500,000) (table 5). This means that to support a 10-percent increase in outputs, smaller farms would require a 5.6-percent increase in all inputs, while larger farms would require an 8-percent increase in inputs. Thus, the greater scale economies available for smaller operations provide a major inducement to increase farm size (compared with the larger farms whose scale elasticities are closer to 1).

However, at the household level, with off-farm income-generating activities included, the scale economies available are lower (scale elasticity is closer to 1; that is, closer to constant returns to scale). Thus, the scale elasticity is higher for all sizes, ranging from 0.73 to 0.96 (table 5). So for smaller farms, a 10-percent increase in all outputs requires a 7.3-percent increase in inputs, while larger farms require a 9.6-percent increase in inputs.[12] More importantly, the difference between the scale elasticities at the household and farm levels is larger for the smaller farms (30 percent) than for the larger farms (around 20 percent). Thus, households operating smaller farms may compensate for the scale disadvantages of their farm business activities with the advantages of off-farm income-generating activities. This advantage may also support the notion that integrated farm and nonfarm labor markets are enabling many small farms to survive and flourish to an extent not thought possible three decades ago (Gardner, 2005).

Off-Farm Work and Economies of Scope

Scope economies measure the cost savings due to simultaneous production of outputs relative to the cost of separate production (see box, p. 14). The concept of economies of scope is useful in assessing the advantages of output diversification. Given the importance of off-farm income to U.S. farm households, scope economies may be expanded to include as output any income-generating activities on or off the farm (household-level scope economies) (see appendix 1).[13] Our estimates for corn and soybean farms show substantial household-level scope economies, 0.24 on average. This means that farm households engaged in off-farm income-generating activities together with the production of traditional farm outputs have cost savings of 24 percent relative to carrying out those activities separately.[14] The cost savings are likely to arise from the sharing of managerial expertise (of the operator and spouse) between onfarm and off-farm activities. Economic evaluations of the farm business alone, then, provide an incomplete view because they exclude off-farm activities, which are an important means of output diversification.

Scope Economies

Scope economies measure the total cost savings due to simultaneous production of outputs relative to the costs of separate production (appendix 1). Given scope economies, it is less costly to produce several outputs in one operation than to produce each output in separate operations (or joint production in one operation generates more output than separate production in two different operations using the same resources). An often-cited example of scope economies is fast food outlets, where savings are obtained by sharing storage, cooking facilities, and customer service in the production of many food products. In general, scope economies may arise from the presence of public inputs or from sharing of imperfectly divisible quasi- fixed inputs in the production of different goods (Fernandez-Cornejo et al., 1992). In our context, farm households achieve scope economies by diversifying or pursuing off-farm activities in addition to the onfarm production of traditional commodities.

To illustrate the possible advantages of "producing" onfarm and off-farm outputs in a farm household, we may use the example of a production possibilities curve (often used in economics). When the production possibilities curve (ACB) is shaped as in the figure, it is advantageous to produce onfarm and off-farm outputs together. As the figure shows, total output produced by a farm household at point C (a combination of onfarm and off-farm outputs) is higher than output produced either at A or B (or a linear combination of both, line AB) while using the same amount of resources.

Scope economies for farm households are likely to arise from the sharing of managerial expertise (and its many components, such as accounting and information processing skills, sales expertise, administrative and technical know-how, etc.) between onfarm and off-farm activities. [15] The expertise of many farm operators and/or their spouses is used in off-farm jobs in private businesses and Government, and in self-employment (Mishra et al., 2002). A USDA survey shows that the largest share of off-farm work done by operators and their spouses is accounted by work in executive, administrative, and managerial positions, service occupations, administrative support, and sales (Covey et al., 2004).

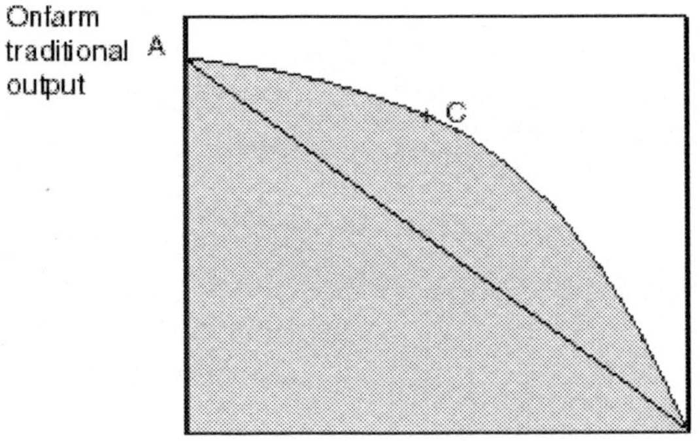

Diagram of a production possibilities curve

Off-Farm Work and Efficiency

Technical efficiency measures how well a farm transforms inputs into outputs given the technology at its disposal (Kumbhakar and Lovell, 2000). Efficiency is of great importance to prevent the waste of resources. Technically inefficient farmers fail to produce the maximum attainable output with the amount of inputs used, and therefore can increase output with the existing level of inputs and available technology.

Two types of technical efficiency are examined here: traditional (farm-level) technical efficiency of the farm business in the production of commodities; and technical efficiency at the household level, which considers both on- and off-farm activities.[16]

Efficiency of the Farm Business

Kumbhakar et al. (1989) examined the effect of off-farm income on farm- level efficiency for dairy farmers. They reasoned that the larger the off-farm component of the operator's income, the less time the operator would spend managing the farm, eroding farm-level efficiency. Indeed, they found that farm-level efficiency of Utah dairy farmers in 1985 was negatively related to off-farm income and that the negative effect was strongest for the smallest farms, which had the largest off-farm incomes.[17] Fernandez-Cornejo (1992) calculated that the farm-level technical efficiency of vegetable farms in Florida was negatively related to off-farm work carried out by the operator. Similar results were obtained by Aigner et al. (2003) for the farm-level efficiency of U.S. corn farmers using 2001 data.

More recently, Goodwin and Mishra (2004) analyzed the relationship between farm-level efficiency and off-farm labor supply. With data collected from 7,699 farms in USDA's 2001 Agricultural Resource Management Survey (ARMS), they used gross cash income (appendix table 1) over total variable costs as an operational measure of farm-level economic efficiency. Greater participation in off-farm labor markets was shown to be significantly associated with lower farm-level efficiency. An additional 1,000 hours engaged in off-farm work would tend to lower the farm-level efficiency ratio by 0.17 with respect to the mean, which was \$1.93 of cash farm income per dollar of variable cost. This effect, while not large, was statistically and economically significant. Such findings support the notion hypothesized by Smith (2002) that off-farm work may hinder "smart farming" and confirm a negative relationship between farming efficiency and the supply of labor to off-farm employment. As theory predicts, more efficient farmers are less likely to work off the farm, reflecting the higher opportunity cost for their labor. Furthermore, the statistical tests performed by Goodwin and Mishra suggest that off-farm labor supply and farm-level efficiency are jointly determined.[18]

Household-Level Efficiency

Rather than estimating the influence of off-farm work on the efficiency of the farm business, we estimated the household-level technical efficiency (including on- and off-farm activities), compared it with farm-level efficiency, and examined how those efficiencies vary with farm size. The technique used in this research isolates the best-practice farm within any size class, and measures technical efficiency by how close other farms are, on average, to the best-practice farms, which are assigned a technical efficiency equal to 1 and said to be on the "frontier." [19]

At the farm level, technical efficiencies of corn/soybean farms increase with size from 0.87 to 0.93 (table 6).[20] However, technical efficiencies at the household level (when off-farm income is included) are higher (around 0.95) and the measures of technical efficiency do not vary across size groups. Moreover, while the beneficial effect of off-farm income occurs at all sizes, it is stronger for smaller farms, whose household-level efficiency levels are comparable with the farm-level efficiencies of the larger farms. This suggests that small corn/soybean farmers have adapted to shortfalls in farm-level efficiency by increasing off-farm income.

Also, the higher household-level efficiencies are consistent with the positive scope economies found. Both findings reflect the more efficient use of resources at the household level, particularly the use of managerial labor (operator and spouse) shared between onfarm and off-farm activities.

Moreover, as Smith (2002) observes, as farm operators and other household members engage in off-farm activities, they have less time available for farm management. This may inhibit their adoption of management-intensive agricultural innovations and lead to less efficient farming.

Off-Farm Work and the Adoption of Agricultural Innovations

Technological change has been acknowledged as a critical component of productivity and economic growth (Solow, 1994; Griliches, 1995). The rapid adoption and diffusion of new technologies in U.S. agriculture has sustained growth in agricultural productivity and ensured an abundance of food and fiber (Huffman and Evenson, 1993). Technological innovations and their adoption have also changed the way farm households regard employment choices (Binswanger, 1974, 1978). Labor-saving technologies, in particular, have allowed farm household members to increase income by seeking off-farm employment (Mishra et al., 2002).[21]

Table 6. Technical efficiency of corn/soybean farms, 1996-2000

		Technical efficiency scores	
Farm type[1]	**Gross sales ($)**	**Farm level (excluding off-farm income)**	**Household level (including off- farm income)**
Farming occupation/	< $100,000	0.87	0.95
lower sales			
Farming occupation/ medium sales	$100,000-$249,999	0.91	0.95
Large family farms	$250,000-$499,999	0.91	0.95
Very large family farms	> $500,000	0.93	0.95
All farms		0.91	0.95

[1]Excluding limited-resource and retirement/residential farms.

Source: Nehring et al., 2005

While profitability (i.e., the extent of yield increases and/or reduction in input costs from an innovation relative to the costs of adoption and current management practices) plays a key role in technology adoption, most studies acknowledge that heterogeneity among farms and farm operators often explains why not all farmers adopt an innovation in the short or long run (Batte and Johnson, 1993; Feder and Umali, 1993; Khanna and Zilberman, 1997; Lowenberg-DeBoer and Swinton, 1997; Rogers, 1961, 1995) (see box, "Factors Influencing Technology Adoption").

Still, assessments of technology adoption using the traditional economic tools pioneered by Griliches (1957) have proven insufficient to explain differing adoption rates for many recent agricultural innovations. The standard measures of farm (accounting) profits, such as net returns (to management), give an incomplete picture of economic returns because they usually exclude the value of management time (Smith, 2002). For example, herbicide-tolerant soybeans were rapidly adopted despite showing no significant advantage in net returns over conventional soybeans. On the other hand, adoption of technologies such as integrated pest management (IPM) has been rather slow despite explicit economic and environmental advantages (Fernandez-Cornejo and McBride, 2002; Smith, 2002). This led to the hypothesis that adoption is driven by "unquantified" advantages, such as simplicity and flexibility, which translate into reduced managerial intensity, freeing time for other uses. An obvious use of managers' time is off-farm employment.

Off-Farm Work as a Factor in Early Studies of Technology Adoption

Early studies of technology adoption viewed off-farm income as influencing adoption of "conservation" practices by providing "supplemental income" to finance conservation expenditures (Blase, 1960). Ervin and Ervin (1982), on the other hand, argued that "off-farm income could reflect the need for supplemental income for family living expenses and essential farm production expenses other than conservation and less time to implement and maintain unfamiliar practices." Survey results on farmers' motivation to seek off-farm income and their view of such employment as permanent rather than temporary, suggest that motivation is closer the view of Ervin and Ervin.

Factors Influencing Technology Adoption

Rural sociologists recognized early that essential differences among farmers can explain why they do not adopt an innovation at the same time. In addition, the characteristics (perceived or real) of an innovation are widely known to influence the adoption decision (Rogers, 1995; Batz et al., 1999). Economists and sociologists have made extensive contributions to the literature on the adoption and diffusion of technological innovations in agriculture (e.g., Griliches, 1957, Feder et al., 1985; Rogers, 1962, 1995). Such research typically focuses on the long-term extent of adoption and the factors that influence the adoption decision.

Farm Structure/Size

A basic hypothesis regarding technology transfer is that the adoption of an innovation will tend to take place earlier on larger farms than on smaller farms. Just et al. (1980) show that, given the uncertainty and the fixed transaction and information costs associated with innovations, there may be a critical lower limit on farm size that prevents smaller farms from adopting. As these costs increase, the critical size also increases. It follows that innovations with large fixed transaction and/or information costs are less likely to be adopted by smaller farms. However, Feder et al. (1985) point out that lumpiness of technology can be offset by the emergence of a service sector (i.e., custom service or consultant). Disentangling farm size from other factors hypothesized to influence technology adoption has been problematic. Feder et al. (1985) caution that farm size may be a surrogate for other factors, such as wealth, risk preferences, and access to credit, scarce inputs, or information. Moreover, access to credit is related to farm size and land tenure because both factors determine the potential collateral available to obtain credit.

Human Capital

The ability to adapt new technologies for use on the farm clearly influences the adoption decision. Most adoption studies attempt to measure this trait through operator age, formal education, or years of farming experience (Fernandez-Cornejo et al., 1994). More years of education and/or experience is often hypothesized to increase the probability of adoption whereas increasing age reduces the probability. Factors inherent in the aging process or the lowered likelihood of payoff from a shortened planning horizon over which expected benefits can accrue would be deterrents to adoption (Barry et al., 1995; Batte and Johnson, 1993). Younger farmers tend to have more education and are often hypothesized to be more willing to innovate.

Risk and Risk Preferences

In agriculture, the notion that technological innovations are perceived to be more risky than traditional practices has received considerable support in the literature. Many researchers argue that the perception of increased risk inhibits adoption (Feder et al., 1985). Hiebert (1974) and Feder and O'Mara (1981) show that uncertainty declines with learning and experience. Innovators and other early adopters are believed to be more inclined to take risks than are the majority of farmers.

Tenure

While several empirical studies support the hypothesis that land ownership encourages adoption, the results are not unanimous and the subject has been widely debated (e.g., Feder et al., 1985). For example, Bultena and Hoiberg (1983) found no support for the hypothesis that land tenure has a significant influence on adoption of conservation tillage. The apparent inconsistencies in the empirical results are due to the nature of the innovation. Land ownership is likely to influence adoption if the innovation requires investments tied to the land. Presumably, tenants are less likely to adopt these types of innovations because the benefits of adoption will not necessarily accrue to them.

Credit Constraint, Location, and Other Factors

Any fixed investment requires the use of own or borrowed capital. Hence, the adoption of a non-divisible technology, which requires a large initial investment, may be hampered by lack of borrowing capacity (El-Osta and Morehart, 1999). Location factors—such as soil fertility, pest infestations, climate, and availability or access to information—can influence the profitability of different technologies across different farms. Heterogeneity of the resource base has been shown to influence technology adoption and profitability (Green et al., 1996; Thrikawala et al., 1999). Irrigation may also influence adoption. Irrigation generally increases yields and profitability and reduces production risk. However, irrigation may also increase risk; for example, it may encourage certain pest populations (Harper and Zilberman, 1989). Contractual arrangements for the production/marketing of the crop are also believed to influence the adoption of certain technologies. Contracts often specify the acreage to be grown or quantity and quality of product to be delivered and may also require the application of certain inputs and practices.

McNamara et al. (1991) used empirical evidence from peanut producers to conclude that integrated pest management (IPM) required substantial time for management and that off-farm employment may present a constraint to IPM participation. Fernandez-Cornejo et al. (1994), Fernandez-Cornejo (1996, 1998), and Fernandez-Cornejo and Jans (1996) found similar results for vegetable and fruit producers. Wozniack (1993) considered livestock feeding innovations and showed that off-farm wage income was inversely related to the likelihood of early adoption and acquiring information.[22] More recent survey results show that operators of high-sales, large, and very large farms—which depend on farm revenues more (and therefore less on off-farm employment) than smaller farms—tend to adopt more management-intensive technologies. For example, around 18 percent of the operators of larger farms adopted precision farming in 1998. In contrast, only 3 percent of the operators of smaller farms (who worked more off-farm hours) adopted precision farming (Hoppe, 2001).

Weaknesses of Early Studies

While insightful, early studies failed to model the interaction of technology adoption and off-farm employment decisions based on the underlying economic theory and consistent with farmers' optimization behavior. Rather, they simply included some measure of off-farm work as one explanatory variable in the adoption decision. Early studies also had some econometric problems, such as not accounting for simultaneity of the off-farm work and adoption decisions and the possibility of self-selection (see appendix 2).[23] Finally, earlier studies did not examine the relationship between technology adoption and household income from farm and off-farm sources.

Modeling the Interaction between Off-Farm Work and Adoption Decisions

To address these issues, we examine the interaction of off-farm income-earning activities and adoption of four agricultural technologies (see box, p. 22) of varying managerial intensity, including herbicide-tolerant crops (Fernandez-Cornejo and Hendricks, 2003; Fernandez-Cornejo et al., 2005), precision agriculture (Fernandez-Cornejo and Southern, 2004), conservation tillage (Fernandez-Cornejo and Gregory, 2004), and Bt (*Bacillus thuringiensis*) corn (Fernandez-Cornejo and Gregory, 2004). We also estimated empirically the relationship between the adoption of these innovations and farm household income from onfarm and off-farm sources.

For this purpose, we expanded the agricultural household model to include the technology adoption decision together with the off-farm work participation decisions by the operator and spouse (appendix 2).[24] We developed an econometric model to examine the interaction of off-farm work and adoption of agricultural technologies, as well as the impact of technology adoption on farm household income (from onfarm and off-farm sources) after controlling for such interaction (appendix 2). The model used data from nationwide surveys of corn and soybean farms in 2000-2001.

We hypothesize that adoption of managerial-saving technologies (such as herbicide-tolerant (HT) soybeans) frees up management time for use elsewhere (notably off-farm employment), leading to higher off-farm income. On the other hand, managerially intensive technologies (such as precision agriculture) would result in less time available for off-farm activities, leading to lower off-farm income.

It is also possible that farmers already working off farm may be more disposed to adopt managerial-saving technologies. This may lead to additional off-farm work and result in even higher off-farm income. Similarly, farmers who are working off farm may be reluctant to adopt managerially intensive technologies.[25]

In either case, we anticipated that adoption of managerial-saving technologies would be associated with higher off-farm income and adoption of managerially intensive technologies would be related to lower off-farm income. (In this report, we show only the empirical validity of the relationship, but not the direction of the causality.)

Table 7. Elasticities of farm household income with respect to the probability of adopting technologies of differing managerial intensity

	With respect to adoption of:			
Elasticity of	**Yield monitors**	**Bt corn[1]**	**Conservation tillage**	**Herbicide-tolerant (HT) soybeans**
Onfarm household annual income	0[2]	0[2]	0[2]	0[2]
Off-farm household annual income	-0.84	0[2]	+0.98	+1.59
Total household annual income	0[2]	0[2]	+0.46	+0.97

[1]Bt = Bacillus thuringiensis)

[2]Statistically insignificant underlying coefficient. The underlying coefficients and their standard errors are shown in appendix 2.

A two-stage econometric estimation method was used to estimate the empirical model (appendix 2). The first stage, the decision model, examines the factors influencing off-farm work participation and technology adoption decisions. The second stage is used to estimate the relationship between technology adoption and household income.

Technology Adoption and Off-Farm Income

We find that the relationship between the adoption of herbicide-tolerant (HT) soybeans and *off-farm* household income is positive and statistically significant (table 7). The elasticity of off-farm household income with respect to the probability of adoption of HT soybeans (calculated at the mean) is,+ 1.59.[26] That is, after controlling for other factors, a 15.9-percent increase in off-farm household income is associated with a 10-percent increase in the probability of adopting HT soybeans. The adoption of HT soybeans is also positively and significantly associated with total household income (from off-farm and onfarm sources). A 9.7-percent increase in total household income is associated with a 10-percent increase in the probability of adopting of HT soybeans. On the other hand, adoption of herbicide- tolerant soybeans did not have a significant relationship with household income from farming (table 7).

Results for adoption of conservation tillage are similar to those obtained for HT soybeans, but of a lesser magnitude (table 7). Controlling for other factors, the association between the adoption of conservation tillage and off-farm household income is positive and statistically significant (elasticity +0.98). An increase in *off-farm* household income of 9.8 percent is associated with a 10-percent increase in the probability of adopting conservation tillage. The association of adoption of conservation tillage and total household income (including both off-farm and onfarm sources) is positive and statistically significant. The elasticity of total household income with respect to the probability of adopting conservation tillage is +0.46.

On the other hand, the relationship between the adoption of yield monitors (an important component of precision agriculture) and off-farm household income is negative and statistically significant (elasticity = -0.84) when we control for other factors. That is, a decrease in off-farm household income by 8.4 percent is associated with a 10-percent increase in the probability of adopting yield monitors. Adoption of yield monitors did not have a statistically significant association with either farm household income or total household income. These results are quite different from those for HT soybeans and conservation tillage. This empirical evidence suggests that yield monitoring techniques are management-intensive compared with the other two technologies, which spare management time.

Finally, the relationship between the adoption of Bt corn with either off- farm or onfarm household income was not statistically significant, indicating that Bt corn may be managerially neutral.

These results are consistent with anecdotal evidence (see box "Selected Agricultural ...") that herbicide-tolerant soybeans save managerial time because of the simplicity and flexibility of weed control. Conservation tillage is also believed to save managerial labor, but to a lesser degree than HT soybeans. Our results for yield monitoring are also consistent with anecdotal evidence that precision farming techniques in general are managerially using. Before the commercial introduction of Bt corn in 1996, most farmers accepted yield losses rather than incur the expense and uncertainty of chemical control. For those farmers, the use of Bt corn was reported to result in yield gains rather than pesticide savings, and savings in managerial time were small.

Selected Agricultural Technologies and Their Managerial Intensity

Herbicide-tolerant (HT) soybeans contain traits that allow them to survive certain herbicides that previously would have destroyed the crop along with the targeted weeds. This allows farmers to use more effective postemergent herbicides, expanding weed management options (Gianessi and Carpenter, 1999). The most common herbicide-tolerant crops are resistant to glyphosate, a herbicide effective on many species of grasses, broadleaf weeds, and sedges. Adoption of HT soybeans has risen rapidly since commercial availability in 1996. HT soybean use rose quickly to about 17 percent of U.S. soybean acreage in 1997 and reached 87 percent in 2005 (FernandezCornejo and McBride, 2002; USDA, NASS, 2003).

Herbicide-tolerant soybeans save managerial time because of the relative simplicity and flexibility of the weed control program. The herbicide-tolerant technology allows growers to apply one herbicide product over the soybean crop at any stage of growth, instead of using several herbicides, to control a wide range of weeds "without sustaining crop injury" (Gianessi and Carpenter, 1999). In addition, using HT soybeans is said to make harvest easier (Duffy, 2001).

Conservation tillage is defined as "any tillage or planting system that maintains at least 30 percent of the soil surface covered by residue after planting" (Conservation Technology Information Center, 2004). It includes no-till, ridge-till, and mulch-till techniques. The impact of conservation tillage in controlling soil erosion and soil degradation is well documented (Edwards, 1995; Sandretto, 1997). By leaving substantial amounts of residue evenly distributed over the soil surface, conservation tillage reduces soil erosion by wind/water, increases water infiltration and moisture retention, and reduces surface sediment and chemical runoff. Adoption of conservation tillage was estimated at 2 percent of planted acreage in 1968 and grew fastest during 1975-85, reaching nearly 28 percent in 1985 (Schertz, 1988). It reached about 37 percent of planted acreage in 2002 (Conservation Technology Information Center, 2004). Conservation tillage is used primarily to grow corn, soybeans, small grains, and cotton.

Conservation tillage is believed to save managerial labor (Sandretto, 1997; USDA, 1998). While it is accepted that adoption of conservation tillage leads to labor savings, its slower rate of adoption compared with HT crops may be because the managerial savings are less.

Bt crops carry the gene from the soil bacterium *Bacillus thuringiensis* (Bt) and are able to produce proteins that are toxic to certain insects. Bt corn, originally developed to control the European corn borer, was planted on 35 percent of corn acreage in 2005, up from 24 percent in 2002. The recent upswing may be due to the commercial introduction in 2003/04 of a new Bt corn variety that is resistant to the corn root-worm.

Before the commercial introduction of Bt corn in 1996, chemical pesticide use was often not profitable to control the European corn borer (ECB) and timely application was difficult (Fernandez-Cornejo and Caswell, 2006). Many farmers accepted yield losses rather than incur the expense and uncertainty of chemical control. For those farmers, the use of Bt corn resulted in yield gains rather than pesticide savings, and managerial time savings were minimal.

Precision agriculture (PA) is often characterized as a suite of technologies used to monitor and manage subfield spatial variability. It includes, for example, global positioning systems, grid soil sampling, yield monitors, and input applicators that can vary rates across a field (Daberkow et al., 2002). These technologies can be used independently or as a package of technologies that includes, for example, the use of grid soil sampling, a variable-rate input applicator, and a yield monitor. PA has been growing relatively slowly. Yield monitors, which provide farmers site-specific data to allow them to vary input application and production practices, are the most extensively adopted PA component. Yield monitors were used in about 33 percent of total corn acreage in 2001 and in about 25 percent of soybean acreage. Adoption of other components of PA is even slower. Adoption of variable-rate applicators reached just 10 percent of corn acreage for fertilizer and 3 percent for pesticides or seeds in 2001 (Daberkow et al., 2002).

Unlike herbicide-tolerant soybeans, which provide savings in management time (and therefore allow operators to obtain higher income from off-farm activities), yield monitors (and precision agriculture in general) are generally believed to be human capital- intensive (Griffin at al., 2004).

CONCLUSIONS

As onfarm and off-farm activities compete for scarce managerial time in U.S. farm operator households, economic decisions (including technology adoption and other production decisions) are likely to shape and be shaped by time allocation within the household. Time allocation decisions are usually not measured directly, but their outcomes, such as onfarm and off- farm income, are observable.

Our research finds that the farm-level efficiency of farm households decreases as off-farm activities increase. Smaller farms, which average the highest off-farm incomes, obtain the lowest farm-level efficiencies. These results support the hypothesis that farm operators who devote more time to off-farm activities have less time to manage the farm. However, examining efficiency from a wider perspective, we find that household-level efficiency (including off-farm income-generating activities) is higher across all farm sizes than farm-level efficiency alone. Moreover, the beneficial effect of off-farm income is higher for smaller farms. In fact, farm households operating small farms achieve efficiency levels comparable to those operating larger farms when off-farm income is included. These results, therefore, suggest that farm households operating small farms have adapted to shortfalls in farming performance by increasing off-farm income.

By including off-farm income-generating activities in the household output portfolio (in addition to the traditional farm products), many farm households, especially those operating smaller farms, are able to enhance diversification. The advantages of such diversification, measured by the household-level economies of scope, are substantial. These results suggest that off-farm employment may enhance onfarm diversification, especially for households operating small farms.

The economic inducement of smaller farms to increase their size (measured by the economic concept of scale economies) is reduced when we include off-farm income. Household-level scale economies (which include off-farm income-generating activities) are closer to constant returns to scale than are farm-level scale economies (which only consider the farm business). However, the beneficial effect of off-farm activities in improving scale economies is more pronounced for households operating smaller farms. These findings provide a different way of measuring the role of off-farm work in improving the economic condition of farm households, particularly those operating small farms.

The adoption of agricultural innovations is also linked to off-farm income through managerial time. For example, the adoption of managerial timesaving technologies is significantly related to higher off-farm household income for U.S. corn/soybean farmers, after controlling for other factors. On the other hand, managerially time-intensive technologies are associated with significantly lower off-farm income.

In a broader sense, these findings confirm the tradeoff between time spent on farm and off-farm activities or, in economic terms, the substitution of economies of scope (derived from engaging in multiple income-generating activities, on and off the farm) for economies of scale.

A number of implications follow. Each of these implications reinforces the importance of understanding farmers' decisions in the context of the farm household rather than the farm operation alone. First, our research provides empirical confirmation of Smith's suggestion that households operating small farms, which lack economies of scale, are more likely to devote time to off-farm opportunities, more likely to adopt management-saving technologies (like herbicide-tolerant crops), and less likely to adopt management-intensive technologies (such as integrated pest management).

The relationship between off-farm work and economic performance also suggests that a farm household's dependence on off-farm income has an effect on the distributional consequences of government policies. Government policies affecting agriculture—such as conservation, research and development, extension, and farm support—may affect farm households differently depending on the relative importance of onfarm and off-farm income-generating activities. Thus, the consequences of government policies depend on the diversity of U.S. farm households, particularly regarding their income sources. For example, a policy promoting the adoption of management-intensive agricultural techniques (such as IPM) may not be fully effective unless it takes into consideration the demands in managerial time imposed by IPM adoption.

This research also has implications for private agricultural research and development (R&D). While innovators often base their economic evaluations of returns to R&D on the expected profitability of potential innovations for farmers (i.e., the extent of yield increases and/or input cost reduction resulting from an innovation relative to the costs of adoption and current management practices), this report shows that there is an important additional element to be included in such evaluations: the value of management time.

REFERENCES

Abler, D. "A Synthesis of Country Reports on Jointness between Commodity and Non-Commodity Outputs in OECD Agriculture." Workshop on Multifunctionality, Directorate for Food, Agriculture, and Fisheries, OECD, Paris, July 2-3, 2001. http://www1.oecd.org /agr/mf/doc /agrmf_abler_rev.pdf#search %20in%20multifunctionality%22

Aigner, D., C.A.K. Lovell, & Schmidt, P. "Formulation and Estimation of Stochastic Frontier Production Function Models," *Journal of Econometrics, 6* (July 1977): 21-37.

Aigner, D.J., Hopkins, J. & Johansson, R. "Beyond Compliance: Sustainable Business Practices and the Bottom Line," *American Journal of Agricultural Economics,* 85(2003): 1126-39.

Ahearn, M.C., & H.S. El-Osta. "The Role of Off-Farm Employment: Permanent or Transitional State." Selected paper presented at the American Agricultural Economics Association (AAEA) annual meetings, Baltimore, MD, Aug. 1993.

Alston, J.M. & Pardey, P.G. *Making Science Pay: The Economics of Agricultural R&D Policy*. Washington, DC: AEI Press, 1996.

Atkridge, J.T. & Hertel, T.W. "Multiproduct Costs Relationships for Retail Fertilizer Plants." *American Journal of Agricultural Economics,* 68(1986): 928938.

Avkiran, N.K. "The Evidence on Efficiency Gains: The Role of Mergers and the Benefits to the Public," *Journal of Banking and Finance*. 23(1999): 991-1013.

Banker, D., & MacDonald. J., editors. *Structural and Financial Characteristics of U.S. Farms: 2004 Family Farm Report*, AIB-797, U.S. Department of Agriculture, March 2005.

Bartlett, P.E. "Part Time Farming: Saving the Farm or Saving Life-Style?" *Rural Sociology* 51 (3)(1986): 289-313.

Barlett, P.F. (1991) "Motivations of Part Time Farmers," in *Multiple Job-Holding Among Farm Families*. M.C. Hallberg, J.L. Findeis, and D.A. Lass (eds.), Iowa State University Press, Ames, IA,.

Barnow, B., Cain, G. & Goldberger, A. (1981). "Issues in the Analysis of Selectivity Bias." E. Stromsdorfer and G. Farkas (eds.) *Evaluation Studies Review Annual*, Vol. 5. Beverly Hills, CA: Sage Publications,.

Barry, P.J., Ellinger, P.N. Hopkins, J.A. & Baker C.B. 1995. *Financial Management in Agriculture, 5th Edition.* Interstate Publishers, Inc. Danville, IL.

Batte, M. & Johnson. R. (1993). "Technology and Its Impact on American Agriculture," in *Size, Structure, and the Changing Face of American Agriculture*, A. Hallum (ed.), Westview Press, Inc.

Battese, G.E., & Coelli, T.J. (1992). "Frontier Production Functions, Technical Efficiency and Panel Data: With Application to Paddy Farmers in India," *Journal of Productivity Analysis, (3)*: 153-169.

Batz, F., Peters K., & Janssen, W. (1999) "The Influence of Technology Characteristics on the Rate and Speed of Adoption," *Agricultural Economics, 21*: 121-130.

Baumol, W.J., Panzar, J.C. & Willig, R.D. (1982.). *Contestable Markets and the Theory of Industry Structure,* New York: Harcourt Brace Jovanovich.

Bender, K., & Hill, L. (2000). *Producer Alternatives in Growing Specialty Corn and Soybeans*. Department of Agricultural and Consumer Economics, University of Illinois, AE-4732.

Berndt, E.R. (1991). *The Practice of Econometrics*. Reading, MA: Addison-Wesley Publishing Co.

Bessant, K.C.((2000). "Part-time Farming Situations Among Manitoba Farm Operators: A Typological Approach," *Canadian Journal of Agricultural Economics, 48(3)*, 259-277.

Blank, S.C. "The Relationship Between Price Risk Management Tools and Off-farm Income." Giannini Foundation of Agricultural Economics. ARER Update 8(6):8-10.

Blank, S., Erickson, Moss, K.W. C.B. & Nehring, R. (2005). "Agricultural Profits and Farm Household Wealth," *American Journal of Agricultural Economics*, *86*(2004), 1299-1307.

Blasé, M.G. (1960). "Soil Erosion Control in Western Iowa: Progress and Problems." Unpublished Ph.D. Dissertation. Iowa State University, Ames, IA.

Boisvert, R. "*Analysis of a Re-Focused Agricultural Policy Within a Farm- Household Framework: Some Data Requirements*," Paper presented at Workshop on the Farm Household-Firm Unit: Its Importance in Agriculture and Implications for Statistics, Wye Campus, Imperial College, University of London. http://household. e%20campus%20london%22 April 12-13, 2002.

Boisvert, R.N., and Chang, H. "*Does Participation in the Conservation Reserve Program and Off-Farm Work Affect the Level and Distribution of Farm Household Incomes*?" Selected Paper prepared for presentation at the American Agricultural Economic Association Annual Meetings, Long Beach, CA, http://agecon.lib.umn.edu/ cgibin/pdf_view.pl?paperid=2 1609&ftype=.pdf July 23-26, 2006.

Bultena, G.L., & Hoiberg E.O. (May-June 1983). "Factors Affecting Farmers' Adoption of Conservation Tillage," *Journal of Soil and Water Conservation*.

Carlton, D.W., & Perloff, J.M. (2000). *Modern Industrial Organization*, 3rd Edition. Reading, MA: Addison-Wesley.

Chavas, Jean-Paul, Ragan Petrie, and Michael Roth. (2005). "Farm Household Production Efficiency: Evidence from Gambia," *American Journal of Agricultural Economics, 87*, 160-179.

Coelli, T. (1996). "A Guide to FRONTIER Version 4.1: A Computer Program for Stochastic Frontier Production and Cost Function Estimation." Mimeo, Department of Econometrics, University of New England, Armidale.

Coelli, T.J., Rao, D.S.P., O'Donnell, C.J. & Battese, G.E. (2005). *An Introduction to Efficiency and Productivity Analysis,* 2nd Edition. New York: Springer.

Conservation Technology Information Center (CTIC). National Crop Residue Management Survey. Conservation Tillage Data. 2004. http://www.ctic. purdue.edu/ CTIC/CRM

Covey, T., Johnson, Morehart J. M., Ryan, J., McGath, C., L. Traub, R. Green, W. McBride, J. Hopkins, R. Strickland, A. Mishra, and K. Erickson. Nov. (2004). *Agricultural Income and Finance Outlook.* U.S. Dept. of Agriculture, Econ. Res. Serv. AIS-82.

Covey, T., Green, R. Jones, C., Johnson J., Morehart, Williams, M. R. McGath, Mishra, C. A., & Strickland, R. (Nov. 2005). *Agricultural Income and Finance Outlook.* U.S. Dept. of Agriculture, Econ. Res. Serv. AIS-83.

Daberkow, S., Fernandez-Cornejo, J. & Padgitt, M. (Nov. 2002). "Precision Ag Adoption Continues to Grow," *Agricultural Outlook*, pp. 24- 27.

Dawson, P., & Hubbard, L. (1985). "Management and Size Economics in the England and Wales Dairy Sector," *Journal of Agricultural Economics, 38*, 27-37.

Deaton, A. (1985). "Panel Data from Time Series Cross-Sections," *Journal of Econometrics, 30*, 109-126.

Duffy, M. "Who Benefits from Biotechnology?" Paper presented at the American Seed Trade Association meeting. Chicago, IL, Dec. 5-7, 2001. http://www.leopold.iastate.edu/pubs/speechwho_benefits_from_biotechnology.pdf#search=%22Duffy%2C%20M.%20%E2%80%9CWho%20Benefits%20from%20Biotechnology%22

Edwards, W.M. (1995)."Effects of Tillage and Residue Management on Water for Crops," in *Crop Residue Management to Reduce Erosion and Improve Soil Quality: Appalachia and Northeast Region.* R.L. Blevins and W.C. Moldenhauer, (eds). Conservation Research Report No. 41. U.S. Dept. of Agriculture, Agricultural Research Service.

El-Osta, H., & Ahearn, M. (1996). *Estimating the Opportunity Cost of Unpaid Farm Labor for U.S. Farm Operators*, U.S. Dept. of Agriculture, Econ. Res. Serv., TB-1848.

El-Osta, H., & Morehart, M. (1999). "Technology Adoption Decisions in Dairy Production and the Role of Herd Expansion," *Agricultural and Resource Economics Review, 28(1)*, 84-95.

Ervin, C.A., & Ervin D.E. (1982). "Factors Affecting the Use of Soil Conservation Practices: Hypothesis, Evidence, and Policy Implications," *Land Economics, 58(3)*, 277-292.

Färe, Rolf, & Primont, D. (1995). *Multi-Output Production and Duality: Theory and Applications*, Kluwer Academic Publishers, Boston/London/Dordrecht.

Feder, G., & Umali, D.L. "The Adoption of Agricultural Innovations: A Review," *Technological Forecasting and Social Change* 43(1993):215- 239.

Feder, G., & O'Mara, G. "Farm Size and the Diffusion of Green Revolution Technology," *Economic Development and Cultural Change* 30(1981):59-76.

Feder, G., Just, R.J. and Zilberman, D. "Adoption of Agricultural Innovations in Developing Countries: A Survey," *Economic Development and Cultural Change* 33(2) (1985):255-98.

Fernandez-Cornejo, J. "Technical Efficiency and Chemical Input Use in Agriculture," *Agricultural and Resource Economics Review* 21 (1992): 11- 21.

Fernandez-Cornejo, J. "The Microeconomic Impact of IPM Adoption." *Agricultural and Resource Economics Review* 25 (1996): 149-60.

Fernandez-Cornejo, J. "Environmental and Economic Consequences of Technology Adoption: IPM in Viticulture," *Agricultural Economics* 18 (1998): 145-55.

Fernandez-Cornejo, J. "Adoption of Genetically Engineered Crops in the U.S." Data product. U.S. Department of Agriculture, Econ. Res. Serv., July, 2005 http://www.ers.usda.gov/Data/BiotechCrops/ .

Fernandez-Cornejo, J., & Hendricks, C. "Off-Farm Work and the Economic Impact of Adopting Herbicide-Tolerant Crops." Selected paper presented at the AAEA meetings, Montreal, Canada, July 27-30, 2003. http://agecon.lib.umn.edu/cgi-bin/pdf_view.pl?paperid=8772&ftype=.pdf

Fernandez-Cornejo, J., & Jans, S. "The Economic Impact of IPM Adoption for Orange Producers in California and Florida," *Acta Horticulturae* 429 (1996): 325-334.

Fernandez-Cornejo, J., & Southern, M. "Managerial Intensity and the Adoption of Agricultural Innovations: The Case of Yield Monitors." Selected Paper presented at the WAEA meetings, San Francisco, CA, July 2004.

Fernandez-Cornejo, J., & Gregory, A. "Managerial Intensity and the Adoption of Conservation Tillage." Selected Paper presented at the NAREA meetings, Annapolis, MD, June 2004.

Fernandez-Cornejo, J., & Caswell, M. *The First Decade of Genetically Engineered Crops in the United States*. EIB-1 1. U.S. Dept. of Agriculture, Econ. Res. Serv., April 2006.

Fernandez-Cornejo, J., & McBride, W.D. *Adoption of Bioengineered Crops*. AER-810. U.S. Dept. of Agriculture, Econ. Res. Serv., May 2002.

Fernandez-Cornejo, J., C. Hendricks, & Mishra, A. "Technology Adoption and Off-farm Household Income: The Case of Herbicide-Tolerant Soybeans," *Journal of Agricultural and Applied Economics* 37(3)(2005): 549-563.

Fernandez-Cornejo, J., Klotz-Ingram, C. and S. Jans. "Farm-Level Effects of Adopting Herbicide-Tolerant Soybeans in the U.S.A.," *Journal of Agricultural and Applied Economics* 34(1)(2002): 149-163.

Fernandez-Cornejo, J., Beach, E.D. & Huang, Wen-Yuan. "The Adoption of IPM Techniques by Vegetable Growers in Florida, Michigan, and Texas," *Journal of Agricultural and Applied Economics* 26(1)(1994):

15 8-72.

Fernandez-Cornejo, J., Daberkow, S. & McBride, W. "Decomposing the Size Effect on the Adoption of Innovations: Agrobiotechnology and Precision Agriculture," *AgBioForum* 4(2)(200 1b): 124-236.

Fernandez Cornejo, J., Stefanou, S.E. Gempesaw, C.M. & Elterich, J.G. "Dynamic Measures of Scope and Scale Economies: An Application to German Agriculture," *American Journal of Agricultural Economics* 74(1992): 329-342.

Ferrier, G.D., & Lovell, C.S.K. (1990). "Measuring Cost Efficiency in Banking: Econometric and Linear Programming Evidence," *Journal of Agricultural Econometrics, 49*, 229-245.

Fuller, A.M. (1991). "Multiple Job-Holding Among Farm Families in Canada," in *Multiple Job Holding Among Farm Families*, M.C. Hallberg, J.L Findeis, and D.A. Lass (eds.). Iowa State University Press, Ames, IA.

Gardner, B.L. "How U.S. Agriculture Learned to Grow: Causes and Consequences." Alan Lloyd Address, Adelaide, Canada. Jan. 2001. http://www.arec.umd.edu/bgardner/papers/Australia

Gardner, B.L. (2002). *American Agriculture in the Twentieth Century*, Harvard University Press: Cambridge, MA.

Gardner, B.L. "The Little Guys Are O.K." (2005). Op-Ed article, *New York Times*, March 7.

Geweke, J., Keane, M. & Runkle, D. (1994) "Alternative Computational Approaches to Inference in the Multinomial Probit Model," *Review of Economics and Statistics, 76(4)*, 609-632.

Gianessi, L.P., & Carpenter, J.E. (1999). *Agricultural Biotechnology: Insect Control Benefits*, National Center for Food and Agricultural Policy, Washington, DC, http://www.bio.org/food

Goodwin, B.K., & Mishra, A.K. (2004). "Farming Efficiency and the Determinants of Multiple Job Holding by Farm Operators," *American Journal of Agricultural Economics, 86,* 722-729.

Goodwin, B.K., & Holt, M.T. (2002). "Parametric and Semiparametric Modeling of the Off-Farm Labor Supply of Agrarian Households in Transition Bulgaria," *American Journal of Agricultural Economics, 84*, 184- 209.

Goodwin, B.K., & Bruer, S.M. (2003). "An Empirical Analysis of Farm Structure and Off-Farm Work Decisions." Selected paper presented at the AAEA meetings, Montreal, Canada, July 27-30.

Gould, B.W., Saupe, W.E. & Klemme, R.M. (1989). "Conservation Tillage: The Role of Farm and Operator Characteristics and the Perception of Soil Erosion," *Land Economics, 65*, 167-82.

Green, G., Sunding, D.. Zilberman, D. & Parker, D. (1996). "Explaining Irrigation Technology Choices: A Microparameter Approach," *American Journal of Agricultural Economics* 78, 1064-72.

Greene, W.H. (1995). *LIMDEP, Version 7.0, User's Manual.* Bellport, NY: Econometric Software.

Greene, W.H. (1997) *Econometric Analysis.* 3rd Edition. Upper Saddle River, NJ: Prentice-Hall.

Greene, W.H. (2000). *Econometric Analysis.* 4th Edition. Upper Saddle River, NJ: Prentice-Hall.

Griliches, Z. (1957). "Hybrid Corn: An Exploration in the Economics of Technological Change," *Econometrica, 25*, 501-22.

Griliches, Z. "R&D and Productivity: Econometric Results and Measurement Issues," in *Handbook of the Economics of Innovation and Technological Change*, P. Stoneman (ed.), Blackwell Publishers Inc., Cambridge, MA.

Griffin, T.W., Lowenberg-DeBoer, J. Lambert, D.M. Peone, J. Payne, T. & Daberkow. S.G. (2004), "Adoption, Profitability, and Making Better Use of Precision Farming Data." Staff Paper #04-06. Department of Agricultural Economics, Purdue University.

Hallam, A. (1991). "Economies of Size and Scale in Agriculture: An Interpretative View of Empirical Measurement, *Review of Agricultural Economics* 13(1)(1991): 155-172.

Hallberg, M.C., Findeis, J.L. & Lass, D.A. (Eds.), *Multiple Job-Holding among Farm Families*, Ames: Iowa State University.

Harper, C.R & Zilberman, D. (1989). "Pest Externalities from Agricultural Inputs, *American Journal of Agricultural Economics,(71)* 692-702.

Heady, E.O. (1952) *Economics of Agricultural Production and Resource Use*, New York: Prentice-Hall.

Heshmati A., & Kumbhakar, S.C. (1992). "Estimation of Technical Efficiency in Swedish Crop Farms: A Pseudo Panel Data Approach," *Journal of Agricultural Economics, 74*, 745-750.

Hiebert, L.D. (1974). "Risk, Learning, and the Adoption of Fertilizer Responsive Seed Varieties," *American Journal of Agricultural Economics, 56*, 764-68.

Hoppe, R.A. *Structural and Financial Characteristics of U.S. Farms: 2001 Family Farm Report*. AIB-768, U.S. Department of Agriculture, Econ. Res. Serv., May 2001.

Hoppe, R.A., Perry J. & Banker, D. (1999) "ERS Farm Typology: Classifying a Diverse Ag. Sector." *Agricultural Outlook*, AGO-266, U.S. Department of Agriculture.

Huffman, W.E. (1980). "Farm and Off-Farm Work Decisions: the Role of Human Capital," *Review of Economics and Statistics, 62(1)*, 14-23.

Huffman, W.E. (1991). "Agricultural Household Models: Survey and Critique," In *Multiple Job Holding Among Farm Families*, M.C. Hallberg, J.L Findeis, and D.A. Lass (eds.). Iowa State University Press, Ames, IA.

Huffman, W.E., & El-Osta, H. (1997). *Off-Farm Work Participation, Off-Farm Labor Supply and Off-Farm Labor Demand of U.S. Farm Operators*. Staff Paper 290, Iowa State University.

Huffman, W.E., & Lange, M.D. (1989). "Off-Farm Work Decisions of Husbands and Wives: Joint Decision Making," *Review of Economics and Statistics, 71*, 471-480.

Huffman,W.E., & Evenson, R.E. (1993). *Science for Agriculture*. Iowa State University Press, Ames, IA.

Just, R., Zilberman, D. & Rausser, G. (1980). "A Putty-Clay Approach to the Distributional Effects of New Technology Under Risk," in *Operations Research in Agriculture and Water Resources*, D. Yaron and C. Tapiero (eds.). New York: North Holland Publishing Company.

Khanna, M., & Zilberman, D. (1997). "Incentives, Precision Technology, and Environmental Protection," *Ecological Economics, 23*, 25-43.

Kimhi, A. (1994). "Quasi Maximum Likelihood Estimation of Multivariate Probit Models: Farm Couples' Labor Participation," *American Journal of Agricultural Economics, 76*, 828-835.

Kimhi, A. (2004). "Family Composition and Off-Farm Participation Decisions in Israeli Farm Households," *American Journal of Agricultural Economics, 86*, 502-5 12.

King, R. (1992). "Management and Financing of Vertical Coordination in Agriculture: An Overview," *American Journal of Agricultural Economics, 74*, 1217-18.

Kinnucan, H., Hatch, U. Molnar, J. & Venkateswaran, M. (1990), "Scale Neutrality of Bovine Somatotropin: Ex Ante Evidence from the Southeast," *Southern Journal of Agricultural Economics, 22*, 1-12.

Kott, P.S. (1998) *Using the Delete-A-Group Jackknife Variance Estimator in NASS Surveys*. RD Research Report No. RD-98-01, U.S. Department of Agriculture, Nat. Agric. Stat. Serv.

Kott, P.S., & Stukel, D.M. (1997). "Can the Jackknife Be Used With a Two-Phase Sample?" *Survey Methodology, 23(2)*, 8 1-89.

Kuchler, F., & Offutt, S. Book review of (1986). "Public Policy and the Changing Structure of American Agriculture," *American Journal of Agricultural Economics, 68(3)*, 764-766.

Kumbhakar, S.C. (1993). "Short-Run Returns to Scale, Farm-Size, and Economic Efficiency," *The Review of Economics and Statistics*, *75(2)*, 336- 341.

Kumbhakar, S.C., & Lovell, C.A.K. (2000) *Stochastic Frontier Analysis*. Cambridge, UK: Cambridge University Press.

Kumbhakar, S.C., Biswas, B. & Von Bailey, D. (1989). "A Study of Economic Efficiency of Utah Dairy Farmers: A System Approach," *The Review of Economics and Statistics, 71 (4)*, 595-604 .

Lass, D.A., & C.M. (1992). "The Supply of Off-Farm Labor: A Random Coefficients Approach," *American Journal of Agricultural Economics, 74*, 400-411.

Lass, D.A., Findeis, J.L. & Hallberg, M.C. (1989), "Off-Farm Labor Employment Decisions by Massachusetts Farm Households," *Northeastern Journal of Agricultural and Resource Economics, 18,* 149-159.

Lass, D.A., Findeis, J.L. & Hallberg, M.C. (1991) "Factors Affecting the Supply of Off-Farm Labor: A Review of the Empirical Evidence." In *Multiple Job Holding Among Farm Families*, M.C. Hallberg, J.L Findeis, and D.A. Lass, eds. Iowa State University Press, Ames.

Leistritz, F.L., Vreugdenhill, H.G. Ekstrom, B.L. & Leholm, A.G. (1985) *Off- Farm Income and Employment of North Dakota Farm Families*. Agricultural Economics Misc. Report No. 88. Agricultural Experiment Station, North Dakota State University, Fargo, ND.

Lopez, R. (1985). "Estimating Labor Supply and Production Decisions of Self Employed Farmers," *American Journal of Agricultural Economics, 84*, 1189-1200.

Lovell, K., and R. Mosheim. (2005). "A Shadow Cost Function Model of the Dairy Sector in the United States." Paper presented at Ninth European Workshop on Efficiency and Productivity, Brussels, Belgium, June 29.

Lowenberg-DeBoer, J., and S. Swinton. (1997) "Economics of Site-Specific Management in Agronomic Crops," in *The State of Site-Specific Management for Agriculture*, F. Pierce and E. Sadler (eds.). ASA Misc. Pub., 369-396, Madison, WI.

Maddala, G.S. (1983). *Limited-Dependent and Qualitative Variables in Econometrics*. Cambridge University Press, Cambridge, UK.

McNamara, K.T., Wetzstein, M.E. & Douce, G.K. (1991). "Factors Affecting Peanut Producer Adoption of Integrated Pest Management," *Review of Agricultural Economics, 13*, 129-139.

Mishra, A.K., & Goodwin. B.K. (1997), "Farm Income Variability and the Supply of Off-Farm Labor," *American Journal of Agricultural Economics, 79*, 880-87.

Mishra, A.K., & Sandretto, C.L. (2002). "Stability of Farm Income and the Role of Nonfarm Income in U.S. Agriculture," *Review of Agricultural Economics, (24)(1)*, 208-221.

Mishra, A.K., El-Osta, H.S. Morehart, M.J. Johnson, J.D. & Hopkins, J.W. (2002). *Income, Wealth, and the Economic Well-Being of Farm Households*. AER-8 12, U.S. Department of Agriculture, Econ. Res. Serv.

Nehring, R., Fernandez-Cornejo, J. & Banker, D. (2005). "Off-Farm Labor and the Structure of U.S. Agriculture: The Case of Corn/Soybean Farms," *Applied Economics, 37*, 633-649.

Newton, D.J. (2005). "Small Farms Can Grow Into Large Enterprises," *Amber Waves*, U.S. Department of Agriculture, Econ. Res. Serv., April. http://www.ers.usda.gov/ Amberwaves/ April05/Findings/SmallFarmsCan Grow.htm

O'Donoghue, E., and R.A. Hoppe. (2005). "Farm Household Income, Farm Structure, and Off-Farm Work," in *Structural and Financial Characteristics of U.S. Farms: 2004 Family Farm Report*, D. Banker and J. MacDonald (eds.). AIB-797, U.S. Department of Agriculture, Econ. Res. Serv., March.

Offutt, S. (2002). "The Future of Farm Policy Analysis: A Household Perspective," *American Journal of Agricultural Economics, 84*, 1189-1200.

Olfert, M.R., (1984). "Off-Farm Labour Supply with Productivity Increases, Peak Period Production and Farm Structure Impacts," *Canadian Journal of Agricultural Economics, 32*, 491-501.

Organization for Economic Cooperation and Development (OECD). *Farm Household Income: Issues and Policy Responses*. 2003. http://www.oecdwash.org/PDFILES/ oecdagr_income6-2-2003.pdf

Panzar, J.C., & Willig, R.D. (1977). "Economies of Scale in MultiOutput Production," *Quarterly Journal of Economics,* 48193.

Panzar, J.C., & Willig, R.D. (1981). "Economies of Scope," *American Economic Review, 71*, 268272.

Paul, C.M., Nehring, R. Banker, D. & Somwaru. A. (2004). "Scale Economies and Efficiency in U.S. Agriculture: Are Traditional Farms History?" *Journal of Productivity Analysis, 22*, 185–205.

Paul, C.J.M., & Nehring. R. (2005). "Product Diversification, Production Systems and Economic Performance in U.S. Agricultural Production, *Journal of Econometrics, 126*, 525-548.

Perry, J., Morehart, M. Banker, D. & Johnson, J. (1997). "Contracting—A Business Option for Many Farmers," *Agricultural Outlook*, U.S. Dept. of Agriculture, Econ. Res. Serv., May.

Pierce, F., & Nowak, P. (1999) "Aspects of Precision Agriculture," *Advances in Agronomy*, Vol. 67, Academic Press.

Rogers, E. (1995) *Diffusion of Innovations*, 1st Edition, 4th Edition. Free Press, New York. 1961.

Sandretto, C. (1997) "Crop Residue Management," *Agricultural Resources and Environmental Indicators*, AH-7 12, U.S. Department of Agriculture, Econ. Res. Serv.

Schertz, D.L. (1988). "Conservation Tillage: An Analysis of Acreage Projections in the United States," *Journal of Soil and Water Conservation* (May-June), 256-58.

Schultz, T.W. "Reflections on Poverty within Agriculture," *Journal of Political Economy,* 63(Feb): 1-15.

Smith, K.R. "Does Off-Farm Work Hinder 'Smart' Farming?" *Agricultural Outlook.* U.S. Dept. of Agriculture, Econ. Res. Serv., AGO-294, (2002), 28- 30.

Singh, I., L. Squire, and J. Strauss (eds.). *Agricultural Household Models: Extensions, Applications, and Policy*. Baltimore: Johns Hopkins University Press, 1986.

Solow, R.M. "Perspectives on Growth Theory," *Journal of Economic Perspectives* 8(1) (1994), 45-54.

Spitze, R.G.F. & Mahoney, R.K. (1991). "Evolving Dimensions of Dual Employment of Illinois Farm Families," in *Multiple Job Holding Among Farm Families*, M.C. Hallberg, J.L Findeis, & Lass, D.A. (eds.). Iowa State University Press, Ames, IA.

Stefanides, Z., & Tauer, L. "The Empirical Impact of Bovine Somatropin on a Group of New York Dairy Farms," (1999). *American Journal of Agricultural Economics, 81*, 95-102.

Sumner, D.A. (1982). "The Off-Farm Labor Supply of Farmers," *American Journal of Agricultural Economics, 64*, 499-509.

Thrikawala, S., Weersink, A. Kachanoski, G. & Fox, G. (1999). "Economic Feasibility of Variable-Rate Technology for Nitrogen on Corn," *American Journal of Agricultural Economics, 81*, 914-27.

Tokle, J. G., & Huffman, W.E. (1991). "Local Economic Conditions and Wage Labor Decisions of Farm and Rural Nonfarm Couples," *American Journal of Agricultural Economics, 73*, 52-70.

United Nations Economic Commission for Europe. *Handbook on Rural Households' Livelihood and Well-Being: Statistics on Rural Development and Agriculture Household Income*. The Wye Group Handbook. Sept. 2005 http://www.unece.org/stats/rural

U.S. Department of Agriculture, Economic Research Service in collaboration with the Natural Resources Conservation Service. *Economic and Environmental Benefits and Costs of Conservation Tillage*. Feb.

U.S. Department of Agriculture, Economic Research Service. (2001). *Food and Agricultural Policy: Taking Stock for the New Century*. http://www.usda.gov/news/ pubs/farmpolicy 01/chapter1.pdf

U.S. Department of Agriculture, Economic Research Service. (2003). *Agricultural Resource Management Survey* (ARMS). http://www.ers.usda.gov/Briefing/ARMS/

U.S. Department of Agriculture, Economic Research Service. (2005). *Farm Structure: Glossary*. http://www.ers.usda.gov/Briefing/FarmStructure/glossary.htm

Varian, H.R. (1992) *Microeconomic Analysis*. Third Edition. New York: W.W. Norton & Company.

Verbeek, M., & Nij man. T. (1993), "Can Cohort Data be Treated as Genuine Panel Data?" *Empirical Economics, 17*, 9-23.

Wozniack, G.D. (1993). "Joint Information Acquisition and New Technology Adoption: Late Versus Early Adoption," *The Review of Economics and Statistics, 75*, 438-445.

Yee, J., Ahearn, M.C. & Huffman, W. (2004). "Links Among Farm Productivity, Off-Farm Work, and Farm Size in the Southwest." *Journal of Agricultural and Applied Economics, 36(3)*, 591-603.

APPENDIX 1—ECONOMIES OF SCALE AND SCOPE AND TECHNICAL EFFICIENCY

This discussion uses two different but related methodologies and data sets and follows the analysis described in Nehring et al., 2005. First, using 1996-2000 survey data, we use an input distance function to estimate scale economies and technical efficiency, and compare these performance measures at the farm and household levels. Second, using 2000 survey data, we set up a multi-activity cost function to analyze labor allocation decisions within the farm operator household and estimate scope economies. We interpret off-farm income-generating activities as outputs, along with corn, soybeans, livestock, and other crops. For both estimations, we use detailed survey information of the farm operator household from USDA's Agricultural Resource Management Survey (ARMS).

Economies of Scale and Technical Efficiency

The analysis of production structure and performance requires representing the underlying multi-dimensional (input and output) production technology. This may be formalized by specifying a transformation function, $\boldsymbol{T(X,Y,R)=0}$, which summarizes the production frontier in terms of an input vector $\boldsymbol{X}$, an output vector $\boldsymbol{Y}$, and a vector of external production determinants $\boldsymbol{R}$. This information on the production technology can also be characterized via an input set, $L(\boldsymbol{Y},\boldsymbol{R})$, representing the set of all X vectors that can produce Y, given the exogenous factors $\boldsymbol{R}$.

An input distance function (denoted by superscript I) identifies the least input use possible for producing the given output vector, defined according to $L(\boldsymbol{Y},\boldsymbol{R})$:

$$D^{I}(\boldsymbol{X},\boldsymbol{Y},\boldsymbol{R}) = Max\{\rho : (x/\rho) \in L(\boldsymbol{Y},\boldsymbol{R})\}. \tag{1}$$

This multi-input, input-requirement function allows for deviations from the frontier. It is also conceptually similar to a cost function, if allocative efficiency is assumed, in the sense that it implies minimum input or resource use for production of a given output vector (and thus, implicitly, costs). However, it does so in a primal/technical optimization or efficiency context, with no economic optimization implied.

For the farm-level model, the $\boldsymbol{Y}$ vector contains Y_1 = crops (corn, soybeans, and other crops), Y_2 = livestock, and, for the household-level model, Y_1 * = crops and livestock, and Y_2* = off-farm income-generating activities, as farm "outputs." With Y_2* included, one might think of $\boldsymbol{Y}$ as a multi-activity rather than a multi-output vector. The components of X are defined as X_1 = land (LD), X_2 = hired labor (L), X_3 = operator labor (including hours worked off-farm)(K), X_4 = spouse labor (including hours worked off-farm) (E), X_5 = capital (F), and X_6 = materials (M).

The scale economies measure may be computed from the estimated model via derivatives or scale elasticities: $-\varepsilon DIY = -\Sigma_m \partial ln\ D^I(\boldsymbol{X},\ Y,\ t)/\partial ln\ Y_m = \varepsilon X1\ Y$ for M outputs $\mathrm{Y_m}$ (similar to the treatment in Baumol et al. (1982) for a multiple-output cost model, and consistent with the output distance function formula in Färe and Primont (1995). However, the inverse measure is more comparable to the cost literature, where the extent of increasing returns or scale economies is implied by the shortfall of the measure from 1. Again, this measure is based on evaluation of (scale) expansion from a given input composition base.

The distance function can be approximated by a translog functional as follows:

$$\ln D^I_{it}(X)_{1'it} = \alpha_0 + \alpha_t t + \Sigma_m \alpha_m \ln X^*_{mit} + 0.5\, \Sigma_m \Sigma_n \beta_{mn} \ln X^*_{mit} \ln X^*_{nit}$$
$$+ \Sigma_k \alpha_k \ln Y_{kit} + 0.5\, \Sigma_k \Sigma_l \alpha_{kl} \ln Y_{kit} \ln Y_{lit} + \Sigma_k \Sigma_m \delta_{km} \ln Y_{kit} \ln X^*_{mit}, \quad (2)$$

or

$$-\ln X_{1'it} = \alpha_0 + \alpha_t t + \Sigma_m \alpha_m \ln X^*_{mit} + 0.5\, \Sigma_m \Sigma_n \beta_{mn} \ln X^*_{mit} \ln X^*_{nit}$$
$$+ \Sigma_k \alpha_k \ln Y_{kit} + 0.5\, \Sigma_k \Sigma_l \alpha_{kl} \ln Y_{kit} \ln Y_{lit} + \Sigma_k \Sigma_m \delta_{km} \ln Y_{kit} \ln X^*_{mit} - \ln D^I_{it}, \quad (3)$$

where i denotes farm and t time period. This functional relationship, which embodies a full set of interactions among the ***X***, ***Y*** and *t* arguments of the distance function, can more succinctly be written as: $-\ln X_{1,it} = TL(\mathbf{X}/X_1, \mathbf{Y}, t) = TL(\mathbf{X^*}, \mathbf{Y}, t)$.

The input distance function is well-suited to measure technical efficiency.

For empirical estimation of technical efficiency, we append a symmetric error term, v, to equation (3) and change the notation "$-\ln D^I_{it}$" to "u." The resulting function (with the subscripts it suppressed for notational simplicity) is: $-\ln X_1 = TL(\mathbf{X^*}, \mathbf{Y}, t) + v - u$, where the term ($-u$) may be interpreted as inefficiency (as technical efficiency measures the distance from the frontier). This method is known as a stochastic frontier production function, where output of a firm is a function of a set of inputs, inefficiency (- u) and a random error v (Aigner et al., 1977; Greene, 1995, 1997, 2000).

To estimate the function, we used Coelli's FRONTIER program (Coelli, 1996), based on the error components model of Battese and Coelli (1992). Since -u represents inefficiency, the technical efficiency scores are given by exp(-u) = $D^I(\mathbf{X^*}, \mathbf{Y}, t)$. If a firm is not technically inefficient (the firm is on the frontier), u is equal to 0 and its technical efficiency score is 1.

In the absence of genuine panel data, repeated cross-sections of data across farm typologies are used to construct a pseudo-panel data set (see Deaton, 1985; Heshmati and Kumbhakar, 1992; Verbeek and Nijman, 1993). The pseudo-panels are created by grouping the individual observations into a number of homogeneous cohorts, demarcated on the basis of their common observable time-invariant characteristics, such as location and ERS farm typology. The subsequent economic analysis then uses the cohort means rather than the individual farm-level observations. ERS farm typology categories are summarized in Nehring et al. (2005). The resulting pseudo panel data set consists of 13 cohorts by State, for 1996-2000, measured as the weighted mean values of the variables to be analyzed. There are a total of 650 annual observations (130 per year), summarizing the activities of 1,934 farms in 1996, 3,890 in 1997, 2,311 in 1998, 3,201 in 1999, and 2,394 in 2000.

Economies of Scope

When a firm produces more than one output, there is a qualitative change in the production structure that makes the concept of economies of scale developed for a single output insufficient. For multiproduct firms, production economies may arise not only because the size of the firm is increased but also due to advantages derived from producing several outputs together rather than separately. Thus, more than one measure is necessary to capture the economies (or diseconomies) related to the scale of operation (volume of output) and the economies related to the scope of the operation (composition of output or product mix). The concepts of economies of scale and scope for multiproduct firms have been developed by Panzar and Willig (1977, 1981) and Baumol et al. (1982). They have been used in agriculture by Akridge and Hertel (1986) and Fernandez-Cornejo et al. (1992).

Economies of scope measure the cost savings due to simultaneous production relative to the cost of separate production. In general, scope economies occur when the cost of producing all products together is lower than producing them separately.

Formally, consider a partition of the output set N into two (disjoint) groups T and N-T. Let Y_T, $Y_{N\text{-}T}$ be the output quantity (subvector) of each of the two groups and Y_N (or simply Y) the output vector, which consists of all the outputs. The respective cost functions $C(Y_T)$, $C(Y_{N\text{-}T})$ give the minimum cost of producing the two output groups separately, and $C(Y_N)$ denotes the minimum cost of producing them together (Nehring et al., 2005).

The degree of economies of scope *(SC)* relative to the (output) set T is defined as:

$$\mathrm{SC} = [C(Y_T)+C(Y_{N\text{-}T}) - C(Y_N)] \,/\, C(Y_N) \tag{1}$$

where SC will be positive if there are economies of scope and negative if there are diseconomies of scope. In our case, we consider the first subset of the partition to include the four conventional outputs (corn, soybeans, other crops, and livestock), $N=\{1,2,3,4\}$ and the second subset the non- conventional off-farm income-generating activities, $N\text{-}T=\{5\}$.

Farms that produce the two output groups separately are those that either produce conventional outputs with no off-farm activities or else those with off-farm work but no conventional outputs. While the sample includes farm households that produce conventional outputs and no off-farm activities, it technically does not include household with zero traditional outputs. However, the sample does include many farm households with very small revenues from traditional outputs because, for statistical purposes, a U.S. farm is currently defined as "any place from which $1,000 or more of agricultural products were sold or normally would have been sold during the year under consideration." (USDA, 2005).

The well-developed restricted cost function is used to estimate the scope economies. Consider n outputs, m variable inputs, and s fixed inputs and other exogenous factors such as location or weather proxies; $\boldsymbol{Y} = (Y1,... Yn)'$ denotes the vector of outputs, $\boldsymbol{X} = (X1,...Xm)'$ denotes the vector of variable inputs, $\boldsymbol{Z} = (Z_1,...Zs)'$ is the vector of non-negative quasi-fixed inputs and other (exogenous) factors, and $\boldsymbol{W} = (W1,... Wm)'$ denotes the price vector of variable inputs. The restricted profit function is defined by:

$$C(W,Y,Z)= Min[W'X : \in T]. \tag{3}$$

Under the usual assumptions on the tech nology (production possibilities set *T*), the restricted cost function is well defined and satisfies the usual regularity conditions.

Using a normalized quadratic variable cost function, which can be viewed as a second-order Taylor series approximation to the true cost function, we obtain:

$$C(W,Y,Z) = a_0 + (a'b'c')\begin{bmatrix} W \\ Y \\ Z \end{bmatrix} = 1/2(W'Y'Z')\begin{bmatrix} B & E & F \\ E & C & G \\ F & G & D \end{bmatrix}\begin{bmatrix} W \\ Y \\ Z \end{bmatrix} \quad (4)$$

where ***W*** is a vector of normalized variable input prices, a_0 is a scalar parameter, and ***a, b,*** and ***c*** are vectors of constants of the same dimension as ***W, Y,*** and ***Z***. The parameter matrices ***B, C,*** and ***H*** are symmetric and of the appropriate dimensions. Similarly, ***E, F ,*** and ***G*** are matrices of unknown parameters.

Using Shephard's lemma, we obtain the demand functions for variable inputs which is estimated together with the cost function. We consider five outputs *Y* (corn, soybeans, other crops, livestock, and operator and spouse off-farm labor), five inputs *X* (hired labor, operator labor, spouse labor, miscellaneous inputs, and pesticides), and use the pesticides price as the *numeraire*. In addition the cost function is specified with two exogenous factors (Nehring et al., 2005).

The normalized quadratic variable cost function and the four cost-share equations are estimated in an iterated seemingly unrelated regression (ITSUR) framework using data for year 2000. The adjusted R^2's were 0.99 for the quadratic cost function, 0.26 for the hired labor input, 0.21 for the operator labor equation, 0.30 for the spouse labor equation, and 0.60 for the miscellaneous inputs equation. However, 48 percent of coefficients for the joint estimates are significant at the 10 percent level.

The own-price effects for the inputs exhibit the expected negative signs. The own-price effect for hired labor is significant at the 10-percent level, while the own-price effects for operator labor and spouse labor are not significant in this cross-section. The own price elasticity of demand for hired labor is highly elastic, with a value of -2.62. In contrast, the own- price elasticities of demand for operator and spouse labor are highly inelastic, with values of -0.105 and -0.283. These results, however, are not directly comparable with cost function studies in the literature that do not include off-farm income- generating activities as an output.

APPENDIX 2—INCORPORATING TECHNOLOGY ADOPTION IN THE FARM HOUSEHOLD MODEL

The Theoretical Framework

This model combines in a single framework the technology adoption and off-farm work decisions by the operator and spouse and follows the analysis developed by Fernandez-Cornejo et al. (2005). The model expands the farm household model offered by Huffman (1991) with several additions to allow for technology adoption. According to the agricultural household model, farm households maximize utility U subject to income, production technology, and time constraints. Household members receive utility from goods purchased for consumption G, leisure (including home time) $L= (L_o, L_s)$ for the operator and the spouse, and from factors exogenous to current household decisions, such as human capital $H = (H_o, H_s)$, and other factors ψ(including household characteristics and weather). Thus:

$$Max\ U= U(G, L, H, \Psi) \tag{1}$$

Subject to the constraints:

$$P_g G = P_q Q - W_x X' + WM' + A \quad \text{(income constraint)} \tag{2}$$

$$Q = Q[X(\Gamma), F(\Gamma), H, \Gamma, R], \ \Gamma \geq 0 \quad \text{(technology constraint)} \tag{3}$$

$$T = F(\Gamma) + M + L, \ M \geq 0 \quad \text{(time constraint)} \tag{4}$$

where P_g and G denote the price and quantity of goods purchased for consumption; P_q and Q represent the price and quantity of farm output; $\boldsymbol{W}_x$ and $\boldsymbol{X}$ are the price and quantity (row) vectors of farm inputs; $\boldsymbol{W} = (W_o, W_s)$ represents off-farm wages paid to the operator and spouse; $\boldsymbol{M} = (M_o, M_s)$ is the amount of time working off-farm by the operator and spouse; $\boldsymbol{F} = (F_o, F_s)$ is the amount of time working on the farm by the operator and spouse; A is other income, including income (from interest, dividends, annuities, private pensions, and rents) and government transfers (such as Social Security, retirement, disability, and unemployment); $\boldsymbol{R}$ is a vector of exogenous factors that shift the production function, and $\boldsymbol{T} = (T_o, T_s)$ denotes the (annual) time endowments for the operator and spouse. The production function is concave and has the usual regularity characteristics. Some technologies offer simplicity and flexibility that translate into reduced management time, freeing time for other uses. In these cases, the amount of time working on the farm by the operator and the spouse $\boldsymbol{F}$ (and possibly the use of other farm inputs $\boldsymbol{X}$) is a function of Γ, the adoption intensity (extent of adoption) of the technology. A technology-constrained measure of (cash) household income is obtained by substituting (3) into (2) (Huffman, 1991):

$$P_g G = P_q Q[X(\Gamma), F(\Gamma), H, \Gamma, R] - W_x X(\Gamma)' + W M' + A \quad (5)$$

The first order conditions for optimality (Kuhn-Tucker conditions) are obtained by maximizing the Lagrangian expression $\mathcal{L}$ over (G, $\boldsymbol{L}$) and minimizing it over the Lagrange multipliers (λ, μ), where $\mu = (\mu_0, \mu_s)$:

$$\mathcal{L} = U(G, L, H, \Psi) + \lambda\ [\ P_q\ Q[\ X(\Gamma), F(\Gamma), H, \Gamma, R] - Wx\ X(\Gamma)'$$

$$+W\ M' + A - P_g\ G\ ] + \mu\ \ [T - F(\Gamma)^2\ M - L] \quad (6)$$

The off-farm participation and adoption decisions may be obtained from the following Kuhn-Tucker conditions:

$$\partial\mathcal{L}/\partial X = \lambda\ (P_q\ \partial Q/\partial X - W_x) = 0 \quad (7)$$

$$\partial\mathcal{L}/\partial F = \lambda\ Pq\ \partial Q/\partial F\ - \mu\ = 0 \quad (8)$$

$$\partial\mathcal{L}/\partial\Gamma = \lambda[P_q[(\partial Q/\partial X)(dX/d\Gamma)'+(\partial Q/\partial F)(dF/d\Gamma)'+\partial Q/d\Gamma]- W_x$$

$$(dX/d\Gamma)'] - \mu\ (dF/d\Gamma)' \leq 0, \quad (9)$$

$$\Gamma \geq 0,\ \ \Gamma \cong \partial\mathcal{L}/\partial\Gamma = 0$$

$$\partial\mathcal{L}/\partial M = \lambda\ W - \ \mu \leq 0,\ \ M \geq 0,\ \ M(\lambda\ W - \ \mu\ ) = 0 \quad (10)$$

$$\partial\mathcal{L}/\partial G = U_G - P_g\ \lambda = 0,\ \partial\mathcal{L}/\partial L = U_L - \mu\ \ = 0 \quad (11\ a,b)$$

$$P_q\ Q[\ X(\Gamma), F(\Gamma), H, \Gamma, R] - W_x\ X(\Gamma)' + WM' + A - P_g\ G = 0 \quad (12)$$

$$T - F(\Gamma) - M - L = 0 \quad (13)$$

where U_L, U_G are the partial derivatives of the function U. Without loss of generality, both the operator and spouse are assumed to have positive optimal hours of leisure and farm work, i.e., equation (8) and (11b) are equalities.

The off-farm participation decision conditions for the operator and the spouse may be obtained from the optimality conditions for off-farm work, equation (10), together with equations (8) and (11b):

$$W \leq \mu/\lambda = P_q \, \partial Q/\partial F \tag{14}$$

where μ/λ is equal to the marginal rate of substitution between leisure and consumption goods (from equations 1 1a and 11b) and $P_q \, \partial Q/\partial F$ represents the value of the marginal product of farm labor for the operator and the spouse. Examining the components of (14), $W_i < \mu_i/\lambda_i$ (strict inequality) indicates that the total time endowment for the operator ($i = o$) or spouse ($i = s$) is allocated between farm work and leisure; optimal hours of off-farm work are zero (corner solution), i.e., $M_i^* = 0$.On the other hand, if $W_i = \mu_i/\lambda$, optimal hours of off-farm work may be positive ($M_i^* > 0$) and $W_i = \mu_i/\lambda = P_q \, \partial Q/\partial F_i$ (interior solution) (Lass et al., 1989; Huffman, 1991; Kimhi, 1994; Huffman and El-Osta, 1997). In this case, the value of the marginal product of farm labor is equal to the off-farm wage rate.[27]

When an interior solution for M occurs, equations (7) and (8) can be solved together, independently of the rest of the Kuhn-Tucker conditions, to obtain the demand functions for onfarm labor, i.e., the optimal production and consumption decisions can be separated since the off-farm wage determines the value of the operator's and spouse's time $(W = \mu/\lambda)$ (Huffman and Lange, 1989; Huffman, 1991).[28]

The demand function for onfarm labor is then $F^*=F(W, W_x, P_q, H, \Gamma, R)$ and the demand for purchased farm inputs $X^* = X(W, W_x, P_q, H, \Gamma, R)$. These optimal input demand functions are substituted in the production function to obtain the supply of farm output $Q^* = S(W, W_x, P_q, H, \Gamma, R)$ and the maximum net household income may be expressed as:

$$NI^* = P_q \, S(W, W_x, P_q, H, \Gamma, R) - W_x X^{*\prime} + WM^\prime + A \tag{15}$$

Solving jointly equations (10), (11), and (15) we obtain the demand for leisure $L^* = L(W, P_g, NI^*, H, \Psi, T)$ for consumption goods $G = G(W, P_g, NI^*, H, \Psi, \Gamma, T)$. The supply function for off-farm time is obtained by substitution of the optimal levels of leisure hours and farm work hours (Huffman, 1991):

$$M^* = T - F^* - L^* = M(W, W_x, P_q, Pg, NI^*, H, \Psi, \Gamma, \xi_i, \Omega, R, T) \tag{16}$$

Finally, a reduced-form expression of total household income is obtained by:

$$NI^* = NI(W_x, P_q, P_g, A, H, \Psi, \Gamma, R, T) \tag{17}$$

As Huffman (1991) notes, when optimal hours of off-farm work hours for the operator or the spouse are zero, the decision process is not recursive and production and consumption decisions must be made jointly. In this case, the arguments for the reduced-form expression of household income are the same as those in (17) but exclude the exogenous variables related to the job characteristics and labor marketability.

The technology adoption decision condition is obtained from the optimality conditions, equation (9) and equations (8) and (11b), noting that the expression in brackets in (9) is the total derivative $dQ/d\Gamma$. Thus, we obtain:

$$Pq\ dQ/d\Gamma - W_x (dX/d\Gamma)' - (\mu/\lambda)(dF/d\Gamma)' \leq 0 \tag{18}$$

But from (11a) and (1 1b) $\mu/\lambda = P_g\ (U_L/U_G)$; then:

$$P_q\ dQ/\Gamma - W_x (dX/d\Gamma)' - P_g (U_L/U_G)(dF/d\Gamma)' \leq 0 \tag{19}$$

The left-hand-side of this expression may be interpreted as the marginal benefit of adoption $P_q\ dQ/d\Gamma$ minus the marginal cost of adoption, which includes the marginal cost of the production inputs $\boldsymbol{W}_x\ (d\boldsymbol{X}/d\Gamma)'$ and the marginal cost ofthe farm work $P_g\ (\boldsymbol{U}_L/U_G)(d\boldsymbol{F}/d\Gamma)'$ (of the operator and the spouse) brought about by adoption (could be negative if adoption saves managerial time), valued at the marginal rate of substitution between leisure and consumption goods (which, when off-farm work hours are positive, equals the off-farm wage rate). It will not be optimal to adopt if the inequality is strict (corner solution), wherein the marginal benefit of adoption falls short of the marginal cost of adoption. An interior solution for the optimal extent of adoption will occur when the equality is strict or when the value of the marginal benefit of adoption is equal to the marginal cost of adoption.

Given the cross-sectional nature of the data, one can use the implicit function theorem to derive expressions for off-farm labor supply for farm operator and spouse and technology adoption (which affects off-farm labor supply of farm operators and spouses) that are functions of wages, prices, human capital, nonlabor income, and other exogenous factors. These factors are replaced in reduced-form representations of labor supply and adoption by observable farm, operator, and houschold characteristics, including human capital. The "ambient variables" (family size, access to urban areas), which might affect the productive capacity of the farm operator and the spouse, are also included. The following section outlines the empirical model and estimation method used to conduct the analysis.

Empirical Model

A two-stage econometric model is specified. The first stage, the decision model, examines the off-farm work participation and the technology adoption decisions. The second stage is used to estimate the impact of adoption on household income.

A simplified "reduced form" approach is followed (Goodwin and Holt, 2002; Goodwin and Mishra, 2004) to specify the empirical model, rather than explicitly estimating a structural model of labor supply. In this approach, the reduced form of the decision model is obtained by specifying the endogenous variables ($\boldsymbol{M}$, $\boldsymbol{F}$, Q_g, $\boldsymbol{X}$) in terms of the exogenous variables, including $\boldsymbol{W}_x$, P_q, P_g, H, Ψ, ξ, Ω, $\boldsymbol{R}$, $\boldsymbol{T}$. Equation (14), implied by the Kuhn-Tucker conditions, is central to the off-farm work decision of the operator and the spouse and equation (19) is central to the adoption decision. Thus, considering a first-order approximation (linear terms) and adding the stochastic terms, the empirical representation of the decision model, which includes the off-farm participation of the operator (20a) and spouse (20b), and the technology adoption decision (20c), is:

$$\beta_o Z_o' + \varepsilon_o \leq 0 \quad (20a)$$

$$\beta_s Z_s' + \varepsilon_s \leq 0 \quad (20b)$$

$$\beta_a Z_a' + \varepsilon_a \leq 0 \quad (20c)$$

where the (row) vectors Z_o, $\boldsymbol{Z_s}$, and $\boldsymbol{Z_a}$ include all the factors or attributes influencing linearly the off-farm participation (operator and spouse) and adoption decisions, and $\boldsymbol{\beta_o}$, $\boldsymbol{\beta_s}$, and $\boldsymbol{\beta_a}$ are vectors of parameters. Assuming that the stochastic disturbances are normally distributed, each of these equations may be estimated by probit. However, because the disturbances $(\varepsilon_o, \varepsilon_s, \varepsilon_a)$ are likely to be correlated, univariate probit equations are not appropriate. Bivariate probit models have been used to model the off-farm employment decisions by the operator and spouse (Huffman and Lange, 1989; Lass et al., 1989; Tokle and Huffman, 1991). Since the decisions to work off farm and the technology adoption decision may be related, all three decisions are modeled together in a multivariate probit model (Greene, 1997). Formally, $[\varepsilon_o, \varepsilon_s, \varepsilon_a]$ ~ trivariate normal (TVN) $[0,0,0;1,1,1;\rho_{12},\rho_{13},\rho_{23}]$, with variances $\rho_{ij}\,(i=j)$ equal to 1 and correlations $\rho_{ij}\,(i \neq j)$ where $i, j = 1,2,3$.

The joint estimation of three or more probit equations was computationally unfeasible until recently because of the difficulty in evaluating high-order multivariate normal integrals. Over the past decade, however, the estimation has been made possible with Monte Carlo simulation techniques (Geweke et al., 1994; Greene, 1997).

The vector $\boldsymbol{Z_i}$ includes (i) farm factors, such as farm size and complexity of the operations; (ii) human capital (operator age/experience and education); (iii) household characteristics (such as the number of children); (iv) off-farm employment opportunities, which will depend on the farms' accessibility to urban areas and the change in the rate of unemployment in nearby urban areas; (v) farm typology; and (vi) government payments.[29] The factors or attributes influencing adoption, included in the vector $\boldsymbol{Z_a}$, are farm factors, human capital, farm typology, a proxy for risk (risk-averse farmers are less likely to adopt agricultural innovations), and crop/seed prices.

The second stage, the income impact model, provides estimates of the impact of adoption on household income after controlling for other factors. The empirical representation of this model—based on equation (17), the reduced-form expression of household income—is $NI^* = NI(W_x, P_q, P_g, A, \boldsymbol{H}, \Psi, \Gamma, \boldsymbol{R}, \boldsymbol{T})$.

After linearizing this reduced form, separating out explicitly the adoption indicator variable, and appending a random disturbance å, assumed to be normally distributed, we have:

$$NI^* = \boldsymbol{\theta} V' + \alpha I + \varepsilon \quad (21)$$

where NI^* represents household income; V is a (row) vector of observable explanatory variables that may influence household income (other than technology adoption) such as prices, human capital, and "ambient variables" (family size, access to urban areas) that may affect the productive capacity of the farm operator and the spouse; I is an indicator variable for adoption ($I=1$ if adoption takes place and $I=0$ otherwise); and θ and α are appropriately dimensioned parameters. The impact of adoption on household income is measured by the estimate of the parameter α. However, as noted by Stefanides and Tauer (1999), if α is to measure the impact of adoption on income of a representative farm, farmers should be randomly assigned among adopter and nonadopter categories. This is not the case, since farmers make the adoption choices themselves. Therefore, adopters and nonadopters may be systematically different and these differences may manifest themselves in farm performance and could be confounded with differences due purely to adoption. This situation, called self-selection, may bias the statistical results unless corrected (Fernandez-Cornejo et al. 2002).

To correct for self-selection bias, we follow Maddala (1983) and Greene (1995) and obtain consistent estimates of the parameters θ and α by regarding self-selection and simultaneity (discussed earlier) as sources of endogenity. Because the dummy variable I cannot be treated as exogenous, instrumental variable techniques are used to purge the dependence of I. The predicted probability of adoption, obtained from the decision model, is used as an instrument for I in equation (21).

Unlike the traditional selectivity model, in which the effects are calculated (separately) using the subsamples of adopters and nonadopters, the impact model uses all the observations and is known as a "treatment effects model," used by Barnow et al. (1981). The treatment effects model consists of the regression $Y = \theta V' + \alpha I + \varepsilon$ where the observed indicator variable I ($I = 1$ if $I^* > 0$ and $I = 0$ if $I^* \leq 0$), indicates the presence or absence of some treatment (adoption of herbicide-tolerant crops in this case) and the unobserved or latent variable I^* is given by $I^* = \delta Z_a' + v_1$ (Greene, 1995).

Total household income (NI^*), as represented in (17), has two components: household income from farming (*FARMHHI*) and off-farm household income (*TOTOFI*). Household income from farming includes farm business household income, operator's paid farm income, household members' paid farm income, etc. (see detailed definitions in appendix table 1). Off-farm household income includes off-farm business income, income from operating other farm businesses, off-farm wages and salaries, etc.

The components of vector V include farm location and typology, operator age, education and experience, number of children, price of soybeans, a measure of specialization on soybean production, a measure of the extent of livestock operations, farm size, and proxies for local labor market conditions.

The data are obtained from the nationwide Agricultural Resource Management Survey (ARMS) developed by USDA (USDA, ERS, 2003). The ARMS survey is designed to link data on the resources used in agricultural production to data on use of technologies, other management techniques, chemical use, yields, and farm financial/economic conditions for selected field crops. The ARMS is a multiframe, probability-based survey in which sample farms are randomly selected from groups of farms stratified by attributes such as economic size, type of production, and land use.

The 2000 data set (used for the HT soybean and Bt corn case study) includes 17 soybean (corn) producing States: Arkansas, Illinois, Indiana, Iowa, Kansas, Kentucky, Louisiana,

Mississippi, Michigan, Minnesota, Missouri, Nebraska, North Carolina, Ohio, South Dakota, Tennessee, and Wisconsin. After selecting those farms that planted soybeans (corn) in 2000 and eliminating those observations with missing data, there were 2,258 observations available for the soybean analysis and 2513 observations for corn.

The 2001 corn data set (used for the yield monitor and conservation tillage case studies) includes observations of 17 corn-producing States. After eliminating observations with missing data, there were 1,763 observations available for analysis.

Because of the complexity of the survey design, a weighted least-squares technique is used to estimate the parameters using full-sample weights developed by the USDA's National Agricultural Statistics Service. Standard errors are estimated using a delete-a-group jackknife method (Kott, 1998; Kott and Stukel, 1997) where a group of observations is deleted in each replication. The sample is partitioned into r groups of observations ($r = 15$) and resampled, thus forming 15 replicates and deleting one group of observations in each replicate.

Appendix table 2 shows the parameter estimates α (equation 21) along with standard errors. These parameters may be interpreted as the derivatives of household income with respect to the probability of adoption and are used to obtain the elasticities shown in table 7.

Appendix Table 1. Household (HH) income variable definitions

1. Household income from farming (FARMHHI) = Farm Business Income HH Share + Operator Paid on Farm + Household Members Paid on Farm + Net Income from Rented Land
Where:
Farm Business Income HH Share = Net Cash Farm Business Income – Depreciation – Gross Income from Rented Land – Operator Paid Onfarm – Income Due to Other Households
Net Cash Farm Income = Gross Cash Farm Income - Cash Operating Expenses
Gross Cash Farm Income = Crop and livestock income including CC loans + Other farm income (includes government payments, income from custom work and machine hire, income from livestock grazing, other farm-related income, income from farm land rented to others, fee income from crops removed under production contract, fee income from livestock removed under production contract).
Total Cash Operating Expenses (hired labor, contract labor, seed, fertilizer, chemicals, fuel, supplies, tractor and other equipment leasing, repairs, custom work, general business, real estate and property taxes, insurance, interest, purchased feed, purchased livestock).
2. Off-Farm Household Income (TOTOFI) = Off-farm business income + Income from operating other farm businesses + Off-farm wages and salaries + Interest and dividend income + Other off-farm income + Rental income
3. Total Household Income (TOTHHI) = Household Income from Farming (FARMHHI) + Off-Farm Household Income (TOTOFI)

Appendix Table 2. Parameter estimates of probability of adoption term of the household income equation for technologies of varying managerial intensity

	Yield monitors			Bt corn			Conservation tillage			Herbicide-tolerant soybean		
	Estimate std. err. t-value			Estimate std. err. t-value			Estimate std. err. t-value			Estimate std. err. t-value		
Onfarm household annual income	25.1	63.8	(0.39)	-13.9	10.9	(–1.29)	6.4	49.5	(0.13)	-30.4	29.8	(-1.02)
Off-farm household annual income	-124.9	35.3	(-3.54)	-36.7	36.2	(–1.07)	87.3	30.3	(2.88)	133.4	67.0	(1.99)
Total household annual income	-100.8	68.7	(-1.47)	-50.6	36.5	(–1.39)	93.9	51.3	(1.83)	104.1	59.0	(1.76)

Note. Standard errors calculated using the delete-a-group jackknife method.

End Notes

[1]Economic researchers have been examining farm economic performance focusing on the farm business for several decades (Heady; Griliches; Dawson and Hubbard; Hallam). Another line of research has focused on the farm household and the labor allocation decisions by the operator and their spouses (Huffman, 1980, 1991; Lass, Findeis, and Hallberg, 1989; Lass and Gempesaw, 1992; Kimhi, 1994, 2004).

[2] Boisvert (2002) stressed not only the growing links between farming activities and off-farm labor markets but also the links between farm household activities, conservation payments, and agricultural pollution.

[3] Loosely, utility is a measure of satisfaction. Economists assume that people act if doing so gives them utility.

[4] The household model initially received a great deal of attention in studies of developing countries' agriculture because of the relative importance of consumption activities in such households. Agricultural economists have also applied these models in developed countries to examine how household members make decisions about the allocation of labor both on and off the farm (Huffman, 1980, 1991; Sumner, 1982; Lopez, 1985; Singh et al., 1986; Lass et al., 1989; Lass and Gempesaw, 1992; Kimhi, 1994, 2004; Mishra and Goodwin, 1997; Goodwin and Holt, 2002). Other analysts have examined income and wealth distributions and links between income instability and consumption/investment (El-Osta and Morehart; Mishra and Morehart). Lopez is one of the few to have considered labor supply and farm production decisions simultaneously. In a very recent application, Chavas et al. used a farm household model to investigate the economic efficiency of farm households in Gambia (Chavas et al., 2005).

[5]Across all farms, operators earned 64 percent of all household off-farm earned income in 2001, spouses earned close to 33 percent, and other members earned 3 percent (O'Donoghue and Hoppe, 2005).

[6]There are, however, some issues regarding the definition of a farm. Since the USDA definition of a farm is not adjusted for inflation, the number of small operations that get defined as farms may increase over time, which may also increase the share of operators working off the farm.

[7]A minority of farmers (18.4 percent of the total in 1987) may be considered as a transitional group, i.e., full-time farmers who worked off farm because they faced heavy losses and high debts. Some of these farmers expected to return to full-time farming when their financial situation was resolved (Bartlett, 1991). Moreover, using agricultural census data spanning 1982 to 1997, ERS researchers identified 644 (out of over 5,000) small part-time farms that managed to grow into large commercial operations. These farms are called emergent adaptive farms (EAF). Off-farm work provided financial support during the early years of the typical EAF, but EAF operators spent more time on farm activities as their businesses expanded: 35 percent of EAF operators worked at least 200 days off the farm in 1987, but that share declined to 16 percent by 1997 (Newton, 2005).

[8] Corn/soybean-producing States are defined as those that account for most of the U.S corn and soybean production. States included are Illinois, Indiana, Iowa, Michigan, Minnesota, Missouri, Nebraska, Ohio, South Dakota, and Wisconsin.

[9] The opportunity cost of labor varies with the farm's region, size, and specialization, the operator's human capital (education and experience); and household characteristics (El–Osta and Ahearn). In addition, opportunity cost estimates may vary with the characteristics of the labor markets, the methodology used, and data sources.

[10] Smaller farms represent a very large share of farm population but a small share of the farm sales. For example, about 44 percent of the farms have sales less than $10,000 and more than 80 percent of the farms have sales below $100,000 (table 3). This distribution, however, is dependent on the definition of farm. In the United States, a farm is currently defined, for statistical purposes, "as any place from which $1,000 or more of agricultural products were sold or normally would have been sold during the year under consideration." (USDA, 2005).

[11] The tradeoff between time spent in onfarm and off-farm activities also manifests itself in Conservation Reserve Program (CRP) participation. Boisvert and Chang (2006) found empirical evidence that a household's decision to participate in the CRP and to work off the farm are made jointly rather than independently. Participation in off-farm work with higher wages provides an incentive for operators to work less on the farm and to take land out of production and commit it to the CRP. As a result, participation in the CRP and off-farm work increase household income.

[12] The scale elasticity increases (moves closer to constant returns to scale) when off-farm income is included because of the theoretical relationship between scale and scope economies in multiproduct firms: "the presence of scope economies 'magnifies' the extent of overall economies of scale beyond what would result from a simple weight sum of product specific economies of scale" (Baumol et al., 1982).

[13] Farms that produce the two output groups separately are those that either produce conventional outputs and no off-farm income or else generate off-farm income but no conventional outputs. While our sample includes farm households that produce conventional outputs with no off-farm activities, it technically does not include households with zero traditional outputs. However, the sample does include many farm households with very small revenues from traditional outputs because, for statistical purposes, in the U.S., a farm is currently defined "as any place from which $1,000 or more of agricultural products were sold or normally would have been sold during the year under consideration" (USDA, 2005). We consider five outputs (corn, soybeans, other crops, livestock, and operator/spouse off-farm labor) and five inputs (hired labor, operator labor, spouse labor, miscellaneous inputs, and pesticides). The method of calculating scope economies, as well as the underlying cost function, is shown in appendix 1.

[14] This result is valid on the average, not necessarily for all the corn/soybean farms studied. For example, it is not likely to be valid for the largest farms in the sample (whose operators are less likely to work off the farm, table 4). As shown in appendix 1, the underlying cost function is a function of the output quantities (and, thus, gross sales), and so are scope economies. The values reported here are calculated at the means of the sample.

[15] As is well known, diminishing marginal labor productivity is a determinant in the allocation of labor between onfarm and off-farm activities. In addition, labor requirements for crop production are often concentrated in very few months of the year. Thus, the marginal productivity of managerial labor for the rest of the year is often very low or negligible (Olfert, 1984).

[16] We have adopted the terminology of "farm-level" and "household-level" efficiency following a recent publication by Chavas et al. (2005). Our earlier terminology (as used in Nehring et al., 2005) was less transparent.

[17] In a subsequent article, Kumbhakar (1993) showed that lower efficiency is the main reason that small farms are less profitable than medium and large farms; another reason being their higher returns to scale (lower scale economies).

[18] There is a two-way relationship between the two variables rather than a cause-and-effect relationship (in economic jargon, each variable is endogenous to the other).

[19] The analysis uses several econometric techniques, including the estimation of an input distance function and stochastic frontier estimation (appendix 1; Nehring et al., 2005) to estimate technical efficiency at the farm (excluding off-farm income-generating activities) and at the household level (including off-farm income-generating activities). Data used were 1995-2003 survey data of corn/soybean farms from 10 States (Illinois, Indiana, Iowa, Michigan, Minnesota, Missouri, Nebraska, Ohio, South Dakota, and Wisconsin), that account for most U.S corn and soybean production.

[20] A farm unit with an efficiency score of 0.8 is said to be 80 percent as efficient as the farms on the 'frontier," i.e., the best performing farms in the data set.

[21] Off-farm employment was also facilitated by economic growth in the nonfarm economy, improved infrastructure (communications and transportation), as well as education level of farm household members (Banker and MacDonald, 2005).

[22] Measures used included off-farm income as a percentage of total household income (Ervin and Ervin, 1982) or days (or hours) per year that the operator worked off-farm (FernandezCornejo, 1996, 1998; FernandezCornejo and Jans, 1996).

[23] Self-selection occurs because farmers are not assigned randomly into groups (e.g., farmers that work off farm or not, adopt or not) but make the choices themselves. Therefore, group members may be systematically different, and these differences may manifest themselves in farm performance and could be confounded with differences due purely to working off farm (or adoption). This situation, called self- selection, may bias the statistical results unless it is corrected (appendix 2).

[24]The agricultural household model (Singh et al., 1986; Huffman, 1980, 1991; Lass et al., 1989; Lass and Gempesaw, 1992; Kimhi, 1994, 2004) combines in a single framework all important economic decisions of the farm household.

[25]As Olfert observes: "Given the nature of nonfarm jobs, where commitments to specific timeframes are frequently more precise than is the case in farming, it is possible that a nonfarm job receives first priority in the allocation of time with farm production undertaken as a second priority." However, Olfert adds: "It may also be the case that the decision regarding time allocation to farm and nonfarm work is made simultaneously or that the off-farm employment decision influences the type and size of farm that is optimal. Farm enterprises that are less demanding in their commitments may be chosen to permit nonfarm employment. Knowing the time commitments required by the nonfarm job, the farm size and type will be organized to accommodate that schedule. Similarly, given the nature of labour requirements on the farm, a choice will be made about the type and amount of nonfarm work."

[26]Results are expressed in terms of elasticity—the percent change in a particular variable (e.g., household income) relative to a small percent change in adoption of the technology from current levels, controlling for other factors. The elasticity results can be viewed in terms of the aggregate change in a particular variable (across an entire agricultural region or sector) relative to aggregate increases in adoption (as more and more producers adopt the technology). However, in terms of a typical farm —that has either adopted or not— the elasticity is usually interpreted as the (marginal) farm-level change associated with an increase in the probability of adoption, away from a given, or current, level of adoption. As shown in appendix 2, the regression model controls for farm location and typology, operator age, education, and experience, number of children, price of the crop, a measure of specialization on soybean/corn production, a measure of the extent of livestock operations, farm size, and proxies for local labor market conditions.

[27] The marginal value of time of the farm operator (or spouse) when all his/her time is allocated to farm work and leisure and none is allocated to off-farm work ($P_q\ \partial Q/\partial F_{i\ |Mi} = 0$) represents the shadow value of farm labor and is called the reservation wage for off-farm work for the operator (i = o) or spouse (i = s). In this context, the operator (or spouse) will work off-farm when his/her reservation wage is less than the anticipated off-farm wage rate and will not work off- farm otherwise. Assuming that both the operator and spouse face wages that are dependent on their marketable human capital characteristics ξi, local labor market conditions, and job characteristics Ω, but not on the amount of off-farm work (Huffman and Lange, 1989; Huffman, 1991; Tokle and Huffman, 1991), the off-farm market labor demand functions are $W_i = W_i\ (4i,\ \xi i, \Omega)$, ($i = o, s$).

[28]Moreover, when an interior solution occurs, from (10), (11a), and (11b) we obtain $U_L/U_G = W/P_g$; that is, the marginal rate of substitution between consumption goods and leisure is equal to the ratio of the wage rate and the price of consumption goods.

[29]Following Goodwin and Holt (2002), some prices are not included in our empirical models since prices are approximately constant across households when data consist of cross- sectional observations taken at a point in time. We did include some prices, like the price of soybeans, but its coefficient was statistically insignificant.

In: America's Family Farms
Editor: Efren J. Tamayo
ISBN: 978-1-60741-751-4

Chapter 9

STRUCTURE AND FINANCES OF U.S. FARMS*

Robert A. Hoppe, Penni Korb, Erik J. O'Donoghue and David E. Banker

ABSTRACT

U.S. farms are diverse, ranging from small retirement and residential farms to enterprises with annual sales in the millions. Nevertheless, most U.S. farms—98 percent in 2004—are family farms. Even the largest farms tend to be family farms. Large-scale family farms and nonfamily farms account for 10 percent of U.S farms, but 75 percent of the value of production. In contrast, small family farms make up most of the U.S. farm count, produce a modest share of farm output, and receive substantial off-farm income. Many farm households have a large net worth, reflecting the land-intensive nature of farming.

Keywords: Contracting, family farms, farm businesses, farm financial performance, farm operator household income, farm operators, farm structure, farm type, million-dollar farms, small farms.

Recommended citation format for this publication:

Hoppe, Robert A., Penni Korb, Erik J. O'Donoghue, and David E. Banker. *Structure and Finances of U.S. Farms: Family Farm Report, 2007 Edition*, EIB-24, U.S. Dept. of Agriculture, Economic Research Service, June 2007.

---

* This is an edited, reformatted and augmented version of a U. S. Department of Agriculture publication dated June 2007.

SUMMARY

U.S. farms are diverse, ranging from very small retirement and residential farms to enterprises with annual sales in the millions of dollars. Farms are operated by individuals on a full- and part-time basis, by multiple generations of a family, by multiple families, and by managers of nonfamily corporations. Some specialize in a single product, while others produce a wide variety of products. Some have full control over their farming processes while others produce commodities under contract to strict specifications. But despite their diversity, most U.S. farms are family farms.

What Is the Issue?

Agricultural policymakers require information on how U.S. farming is organized. USDA's Economic Research Service (ERS) produces a periodic report with that information. The *Family Farm Report, 2007 Edition* is the most recent in the series, providing agricultural policymakers with an accurate, detailed, and unbiased source of information on how farming in the United States is organized, including the relationship of farm size and type to agricultural production, financial performance, sources of farm household income, and the extent of off-farm work. The report provides a sense of the financial position of family farms in general and for different types of family farms.

What Are the Major Findings?

Most U.S. farms—98 percent in 2004—are family farms, defined as operations organized as proprietorships, partnerships, or family corporations that do not have hired managers. Nonfamily corporations make up a small and stable share of farm numbers and sales, accounting for less than 1 percent of farms and 6-7 percent of farm product sales in each agricultural census since 1978.

Small family farms account for most U.S. farms and farm assets

Small family farms (sales less than $250,000) accounted for 90 percent of U.S. farms in 2004. They also held about 68 percent of all farm assets, including 61 percent of the land owned by farms. As custodians of the bulk of farm assets—including land—small farms have a large role in natural resource and environmental policy. Small farms accounted for 82 percent of the land enrolled by farmers in the Conservation Reserve and Wetlands Reserve Programs (CRP and WRP).

Large-scale family farms and nonfamily farms produce the largest share of agricultural output

Large-scale family farms, plus nonfamily farms, made up only 10 percent of U.S. farms in 2004, but accounted for 75 percent of the value of production. Nevertheless, small farms made significant contributions to the production of specific commodities, including hay, tobacco, wheat, corn, soybeans, and beef cattle.

Distribution of farms, total production, and assets, 2004

Farm type	Farms	Value of production	Farm assets
	Percent of U.S. total		
Small family farms:[1]			
Limited-resource	9.4	1.0	5.5
Retirement	16.1	2.0	11.3
Residential/lifestyle	39.7	5.3	23.7
Farming-occupation			
Low-sales	18.8	5.5	16.9
Medium-sales	6.3	10.8	10.3
Large-scale family farms:[1]			
Large family farms	4.1	14.8	9.1
Very large family farms	3.4	45.4	16.1
Nonfamily farms[1, 2]	2.2	15.2	7.1

[1]Small farms have sales less than $250,000; large-scale farms have sales of $250,000 or more; no sales limit for nonfamily farms.

[2]Nonfamily farms include those organized as nonfamily corporations or cooperatives, as well as any other farms operated by hired managers. Also includes farms held in estates or trusts.

Source: USDA, ERS, 2004 Agricultural Resource Management Survey, Phase III.

The number of larger farms is growing

The number of farms with sales of $250,000 or more grew steadily between the 1982 and 2002 Censuses of Agriculture, with sales measured in constant 2002 dollars. The growth in the number of these larger farms was accompanied by a shift in sales in the same direction. The most rapid growth was for farms with sales of $1 million or more. By 2002, million-dollar farms alone accounted for 48 percent of sales, compared with 23 percent in 1982.

For the most part, large-scale farms are more viable businesses than small family farms

The average operating profit margin and rates of return on assets and equity for large and very large family farms were all positive in 2004, and most of these farms had a positive operating profit margin. Small farms were less viable as businesses. Their average operating profit margin and rates of return on assets and equity were negative. Nevertheless, some farms in each small farm group had an operating margin of at least 20 percent. In addition, a majority of each small farm type had a positive net farm income, although the average net income for each small-farm type was low compared with large-scale farms.

Small farm households rely on off-farm income

Small farm households typically receive substantial off-farm income and do not rely primarily on their farms for their livelihood. Most of their off-farm income is from wageand-salary jobs or self-employment. Because of their off-farm work, small farm households are affected significantly by the nonfarm economy. Households operating retirement or limited-resource farms, however, receive well over half of their income from such sources as Social Security, pensions, dividends, interest, and rent, reflecting the ages of operators on such farms.

Payments from commodity-related programs and conservation programs go to different types of farms

The distribution of commodity-related program payments is roughly proportional to the harvested acres of program commodities. As a result, medium-sales ($ 100,000-$249,999) and large-scale farms received 78 percent of commodity-related government payments in 2004. In contrast, CRP, which pays the bulk of environmental payments, targets environmentally sensitive land rather than commodity production. Retirement, residential/lifestyle, and low-sales small farms received 62 percent of conservation program payments in 2004. However, most farms—61 percent in 2004—receive no government payments and are not directly affected by farm program payments.

A growing number of farms operate under production and marketing contracts to guarantee an outlet for their production

About two-fifths of U.S. agricultural production is produced or marketed under contract, although the share varies by commodity and type of farm. Relatively few small family farms use production and marketing contracts, while 64 percent of very large family farms use contracts and, as a group, produce 61 percent of the value of production grown under contract.

How Was the Study Conducted?

The 2004 Agricultural Resource Management Survey (ARMS) is the main source of data in the *Family Farm Report, 2007 Edition*. ARMS is an annual survey designed and conducted by ERS and USDA's National Agricultural Statistics Service (NASS). Various censuses of agriculture, ERS estimates of farm productivity, NASS estimates of the number of farms, and labor force data from the Bureau of Labor Statistics are also used in this report, particularly for long-term trends. The report uses the farm classification system (see table, p. iii) developed by ERS to examine farm structure in the United States.

INTRODUCTION

Farming in the United States is very diverse, ranging from very small retirement and residential farms to enterprises with annual sales in the millions of dollars. Farms are operated by individuals on a part-time basis, by multiple generations of a family, and by managers of nonfamily corporations. Some specialize in a single product; others produce a wide variety of products.

The *Family Farm Report, 2007 Edition* presents comprehensive information about the structure and finances of the various types of family farms in the United States. This report covers the following aspects of farm structure:

The number, size distribution, and tenure of U.S. farms.
The specialization and diversification of farms.
Farm operator demographics, including age, education, gender, and race/ethnic origin.

The sources and levels of operator household income and wealth.

The share of farms receiving government payments and the distribution of government payments by type of farm.

The business organization of farms—whether they are organized as proprietorships, partnerships, or corporations.

ERS developed a farm classification (see box, "Farm Types, 2004") to group farms into more homogeneous categories, based primarily on annual gross sales of the farm and major occupation of the operator. By using these homogeneous groups in this report, a clearer picture emerges of the status of farms in the United States today.

As in recent years, the Agricultural Resource Management Survey (ARMS)—an annual survey—is the main source of data in the *Family Farm Report, 2007 Edition*. The ARMS is jointly designed and conducted by USDA's Economic Research Service (ERS) and National Agricultural Statistics Service (NASS).[1] The report also draws on various censuses of agriculture, ERS estimates of farm productivity, NASS annual estimates of the number of farms, and Bureau of Labor Statistics (BLS) labor force data. These additional sources of data are particularly useful when following trends over long periods of time.

This report depicts farm structure and financial status as of 2004, the most recent year for which ARMS data were available at the time of writing, and 2004 was atypical year for farming farm income was $83 billion in 2004 (figure 1), much higher than in 2003 ($61 billion), the annual average during the previous 10 years ($55 billion), and the previous peak in 1996 ($69 billion). Net farm income is expressed in 2004 dollars here, using the GDP chain-type price index to adjust for price changes.

FARM TYPES, 2004

This farm classification focuses on the "family farm" or any farm organized as a sole proprietorship, partnership, or family corporation. Family farms exclude farms organized as nonfamily corporations or cooperatives, as well as farms with hired managers

SMALL FAMILY FARMS (GROSS SALES LESS THAN $250,000)[1]

Limited-resource farms. Farms with gross sales less than $100,000 in 2003 and less than $105,000 in 2004.[2] Operators of limited-resource farms must also receive low household income in both 2003 and 2004. Household income is considered low in a given year if it is less than the poverty level for a family of four, or it is less than half the county median household income. Operators may report any major occupation except hired manager.

Retirement farms. Farms whose operators report they are retired.[3]

Residential/lifestyle farms. Farms whose operators report a major occupation other than farming.[3]

Farming-occupation farms. Farms whose operators report farming as their major occupation.[3]

Low-sales farms. Gross sales less than $100,000.

Medium-sales farms.[4] Gross sales between $100,000 and $249,999.

LARGE-SCALE FAMILY FARMS (GROSS SALES OF $250,000 OR MORE)

Large family farms. Gross sales between $250,000 and $499,999.
Very large family farms. Gross sales of $500,000 or more.

NONFAMILY FARMS

Nonfamily farms. Farms organized as nonfamily corporations or cooperatives, as well as farms operated by hired managers. Also includes farms held in estates or trusts.

Note: A farm is defined as any place that produced and sold—or normally would have produced and sold—$1,000 worth of agricultural products during a given year (USDA, NASS, 2005, p. 3-1).
[1]The National Commission on Small Farms selected $250,000 in gross sales as the cutoff between small and large farms (U.S. Dept. Agr., Nat'l. Comm. on Small Farms, 1998, p. 28).
[2]The original gross sales cutoff was established at $100,000 for 2003. The cutoff for subsequent years is adjusted by the index of prices paid by farmers.
[3]Excludes limited-resource farms whose operators report this occupation.
[4]This type was called "high-sales" farms in earlier publications.

The *Family Farm Report* series features a special topic each year, starting with the previous edition of the report (Hoppe and Banker, 2006). The special feature this year is "The Shift to Larger Farms," which examines changes in the distribution of farm and gross farm sales (by constant dollar sales classes) between the 1982 and 2002 Censuses of Agriculture.

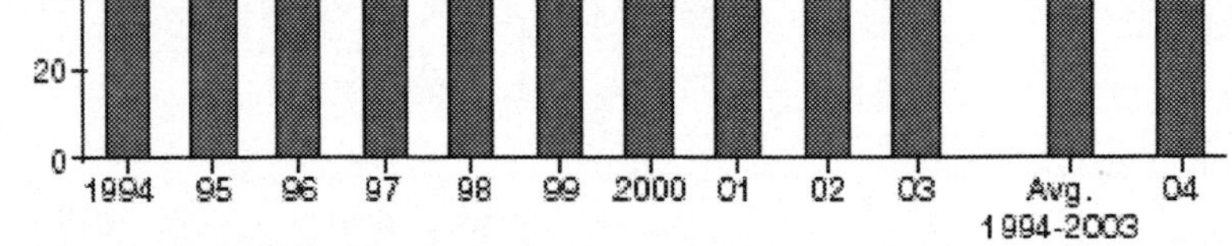

[1]Deflated with the GDP chain-type price index.
Source: USDA, Economic Research Service, U.S. and State Farm Income Data.

Figure 1. Real net farm income, 1994 to 2004
In 2004, net farm income was 50 percent higher than the average for the previous 10 years

U.S. FARMS: NUMBERS, SIZE, AND OWNERSHIP

In the 1930s, two important longrun trends began that affected the number of U.S. farms. First, nonagricultural employment resumed growing after the 1933 low point of the Great Depression (figure 2). Second, farm productivity began to increase steadily (figure 3), starting about 1937 (Cochrane, 1993, pp. 360-363). Productivity growth led to excess capacity in agriculture, farm consolidation, and farm operators and laborers leaving farming to work in the growing nonfarm economy. The decline in farm numbers slowed in the 1980s and nearly stopped in the 1990s. By 2005, about 2.1 million farms remained, and less than 2 percent of U.S. workers were employed in agriculture. The remaining farms, however, vary in size and their share of total production.

Share of Farms, Production, and Assets

Ninety-eight percent of U.S. farms are family farms. The remaining 2 percent are nonfamily farms, which produce 15 percent of the value of agricultural output (figure 4).[2] Two features of family farms stand out. First, there are many small family farms (< $250,000 annual sales), making up 90 percent of all U.S. farms. Second, large-scale family farms account for 60 percent of all production.

Nevertheless, small farms make significant contributions to the value of production for specific commodities (figure 5), including wheat, corn, soybeans, hay, tobacco, beef, and "other livestock." At the other extreme, small farms contribute a minuscule share to the value of production for hogs and poultry. The largest share of small farm production occurs among medium-sales farms ($ 100,000-$249,999), which account for 11 percent of total U.S. production.

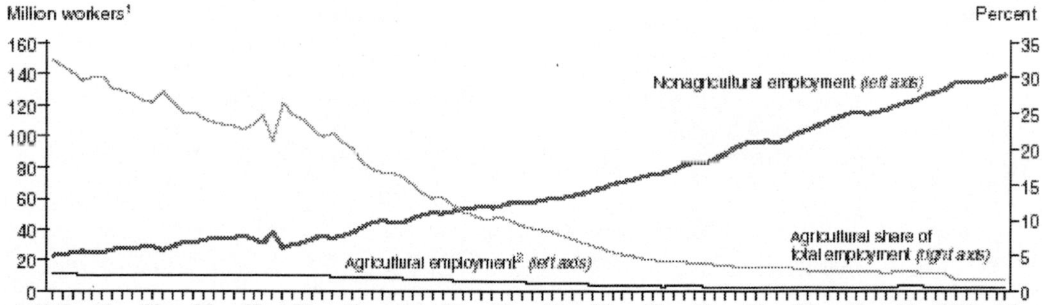

[1]Persons at least 14 years old prior to 1947; persons at least 16 years old in 1947 and later years.

[2]From 2000 onward, estimates of agricultural employment actually are for "agricultural and related industries." For more information, see the U.S. Department of Labor, Bureau of Labor Statistics (2003, p. 20).

Source: USDA, Economic Research Service, compiled from Bureau of Labor Statistics data (U.S. Office of the President, 2006, pp. 324-325; U.S. Department of Commerce, 1975, p. 126).

Figure 2. Agricultural and nonagricultural employment, 1910 to 2005
Agriculture's share of total U.S. ermployment has fallen to less than 2 percent

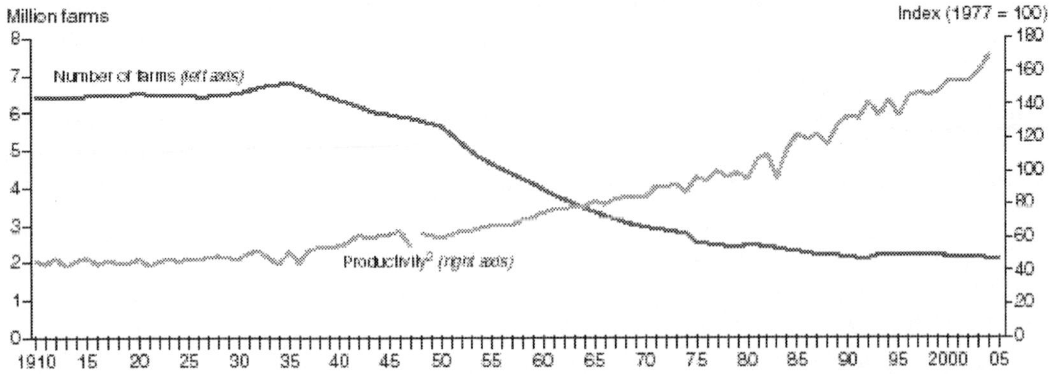

[1]Farm output per unit of total factor input (total factor productivity), available through 2004.

[2]The break in the productivity line reflects the introduction of new methodology beginning with the 1948 estimate. The new methods had minor impacts on the estimates. For more information, see Ahearn et al. (1998, pp. 15-21).

Source: USDA, Economic Research Service, compiled from National Agricultural Statistics Service annual estimates of the number of farms from the June Agricultural Survey and from ERS estimates of farm productivity. ERS productivity indices prior to 1948 came from Johnson (1990).

Figure 3. Farm productivity[1] and number of farms, 1910 to 2005
The number of farms declined as productivity increased

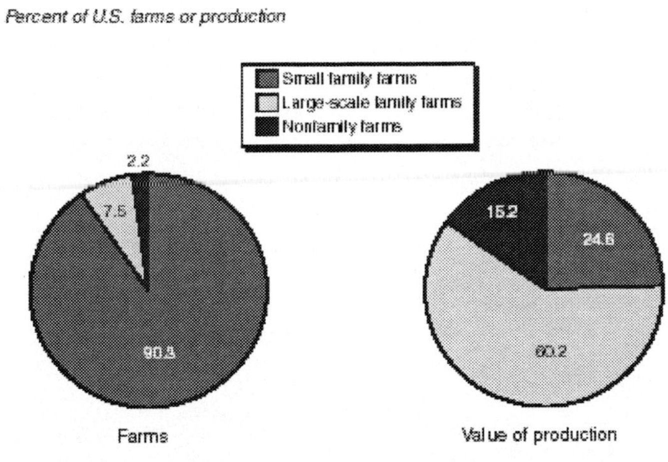

Source: USDA, Economic Research Service, 2004 Agricultural Resource Management Survey, Phase III.

Figure 4. Share of total farms and value of production, 2004
Large-scale family farms account for 60 percent of production

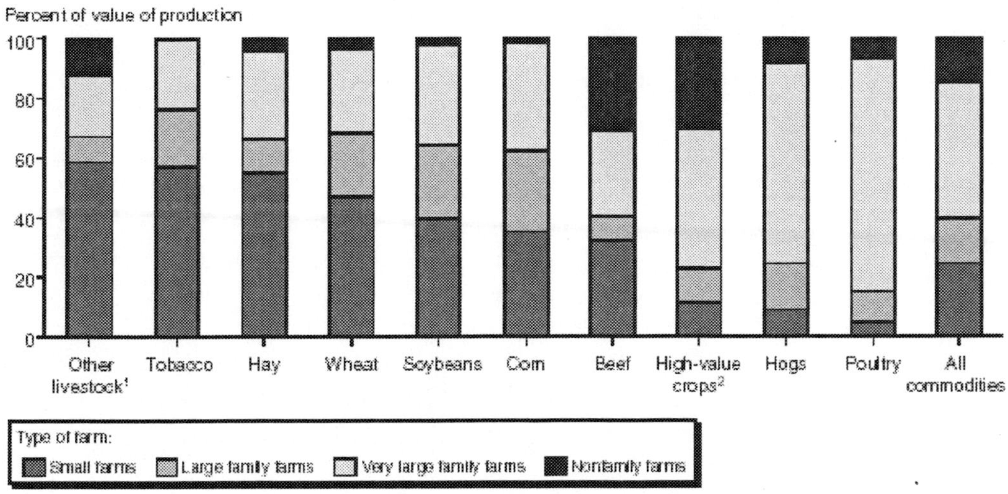

[1]Sheep, lambs, wool, goats, goats' milk, mohair, horses, ponies, mules, donkeys, bees, honey, aquaculture, mink, rabbits, other fur bearing animals, bison, deer, elk, llamas, etc.

[2]Vegetables, fruits and tree nuts, and nursery and greenhouse products.

Source: USDA, Economic Research Service, 2004 Agricultural Resource Management Survey, Phase III.

Figure 5. Distribution of the value of production for selected commodities, 2004
Small farms produce a substantial share of several commodities

The share of assets and land held by small farms is substantially more than indicated by their 25-percent share of production. Small farms hold about 68 percent of all farm assets, including 61 percent of the land owned by farms (figure 6). Because of their large land holdings—in aggregate—small farms are important in conservation efforts. Small farms account for 82 percent of the land farmers enrolled in the Conservation Reserve Program (CRP) and Wetlands Reserve Program (WRP).

Farm Size and Tenure

Variation in size—measured in sales, acres, and labor use—helps explain the distribution of agricultural production. The 1.4 million limited-resource, retirement, and residential/lifestyle farms account for only 8 percent of production because most of these farms are very small (table 1). Roughly three-fourths of the farms in each of the three groups have annual gross sales of less than $10,000. The average acreage operated for farms in these three groups is also small, ranging from 163 to 212 acres.

Median Acres Operated

Average (or mean) acreage operated may not best indicate the size of a typical farm in a group because a few high-acreage farms may raise the average well above the acreage operated on most farms. Median acreage operated—the midpoint of the distribution of farms by acres operated—is a better indicator. Median acreage operated ranges from 60 to 80 acres for limited-resource, retirement, and residential/lifestyle farms, which means the typical farm in each of these groups is even smaller than suggested by the group's average acreage.

Although only half of low-sales farms (< $100,000) have gross sales of less than $10,000, three-fourths have gross sales of less than $50,000. Median acres operated was 145 acres per low-sales farm, roughly double the medians for limited-resource, retirement, or residential/lifestyle farms. Median acreage is much larger for medium-sales small farms and large- scale farms, ranging from 530 to 1,055 acres.

The high average acreage for nonfamily farms, more than 1,200 acres, reflects a small share of farms in the group with very large acreages. In contrast, the median is just 173 acres, which is more consistent with the 40- percent share of nonfamily farms with gross sales less than $10,000. Most of these very small nonfamily farms (77 percent) are classified in the "other" category of business organization, which includes farms in estates or trusts and farms organized as cooperatives. Very small nonfamily farms in the other organization category are more likely to be in estates and trusts than to be cooperatives. (Business organization is discussed in more detail later in this report.)

Million-dollar Farms

Forty-two percent of very large family farms ($500,000 or more annual sales) and 9 percent of nonfamily farms have gross sales of at least $1 million. These "million-dollar" farms make up less than 2 percent of all U.S. farms, but they account for 45 percent of the value of production. The number of million-dollar farms increased by 22 percent between 2003 and 2004, from 28,300 to 34,500, as 2004 was a very good year for the farm sector.

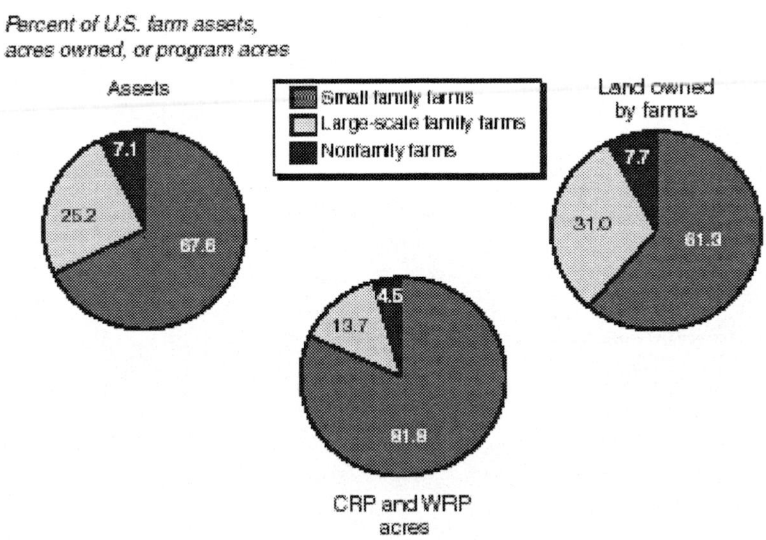

Source: USDA, Economic Research Service, 2004 Agricultural Resource Management Survey, Phase III.

Figure 6. Share of farm business assets, acres owned, and acres enrolled in the Conservation Reserve Program and Wetlands Reserve Program, 2004
Small farms account for most farm assets

Table 1. Farm size, tenure, and labor, by farm type, 2004

Item	Small family farms					Large-scale family farms		Nonfamily farms	All farms
				Farming-occupation					
	Limited-resource	Retire-ment	Residential /lifestyle	Low-sales	Medium-sales	Large	Very large		
	Number								
Total farms	197,734	338,671	837,542	395,781	133,299	86,087	71,708	47,103	2,107,925
	Percent of U.S. total								
Distribution of:									
Farms	9.4	16.1	39.7	18.8	6.3	4.1	3.4	2.2	100.0
Value of production	1.0	2.0	5.3	5.5	10.8	14.8	45.4	15.2	100.0
	Percent of group								
Sales class:									
Less than $10,000	76.2	72.6	71.6	47.1	na	na	na	39.5	57.0
$10,000 to $49,999	19.2	18.6	20.9	29.4	na	na	na	18.1	19.0
$50,000 to $99,999	4.0	6.3	5.2	23.5	na	na	na	*9.6	8.1
$100,000 to $174,999	d	*1.9	1.8	na	59.4	na	na	*6.8	5.0
$175,000 to $249,999	na	*0.6	0.6	na	40.6	na	na	*4.1	3.0
$250,000 to $499,999	na	na	na	na	na	100.0	na	5.9	4.2
$500,000 to $999,999	na	na	na	na	na	na	57.8	7.0	2.1
$1,000,000 or more	na	na	na	na	na	na	42.2	8.9	1.6
	Acres per farm								
Acres operated:									
Mean (average)	167	212	163	413	*1,170	1,700	*3,138	1,232	470
Median[1]	60	80	67	145	530	834	1,055	173	100
	Annual person equivalents of labor per farm								
Average person equivalents of labor[2, 3]	0.997	0.928	0.782	1.500	2.580	3.265	8.156	5.385	1.529

Table 1. (Continued)

		Small family farms					Large-scale family farms		
				Farming-occupation					
	Limited-resource	Retire-ment	Residential /lifestyle	Low-sales	Medium-sales	Large	Very large	Nonfamily farms	All farms
	Percent of total hours								
Share of hours worked by:[4]									
Principal operator[3]	71.7	67.0	63.5	65.4	59.7	47.5	19.2	19.1	51.4
Spouse[3]	12.2	17.2	18.6	16.1	12.2	10.9	4.5	*2.1	12.4
Hired labor	2.5	2.2	2.5	5.1	#10.8	21.8	55.5	61.6	19.8
	Percent of group								
Tenure:									
Full owner	68.3	79.5	67.2	60.3	20.5	19.1	26.1	72.1	61.8
Part owner	25.7	*19.1	27.6	32.8	68.1	66.5	60.3	18.4	32.1
Tenant[5]	*6.0	*1.3	5.2	6.9	#11.4	14.3	13.6	9.5	6.1

d = Data suppressed due to insufficient observations.

na = Not applicable.

* = Standard error is between 25 percent and 50 percent of the estimate.

= Standard error is greater than 75 percent of the estimate.

[1]Midpoint of the distribution of farms by acres operated. Half the farms in a group operate more acres than the median, while the other half operate fewer acres than the median.

[2]One annual person equivalent equals 2,000 hours of labor, or 50 weeks per year times 40 hours per week.

[3]Includes paid and unpaid hours.

[4]Shares worked by other operators, unpaid workers, and contract labor are not shown separately.

[5]Farms that rent all the land they operate. Also includes farms owning less than 1 percent of the land they operate.

Source: USDA, Economic Research Service, 2004 Agricultural Resource Management Survey, Phase III.

Labor hours[3]

One measure of annual labor use is the "person equivalent," defined as 2,000 hours, or 40 hours of work per week for 50 weeks per year. Residential/lifestyle, retirement, and limited-resource farms use the least labor of all the farm types, 1 person equivalent or less. Labor use jumps to 1.5 person equivalents for low-sales farms and increases with sales to 8.2 person equivalents for very large farms. Nonfamily farms use 5.4 person equivalents, on average. This estimate, however, reflects heavy labor use by relatively few farms. Only 20 percent of nonfamily farms use more than 5 person equivalents of labor, while 46 percent use less than 1.

The labor used on farms can come from a variety of sources: operators and their spouses, secondary operators, unpaid workers, hired labor, and contract labor. Nevertheless, operators are a significant source of labor for most farm types. The operator provides 60 to 70 percent of the labor for each type of small farm, and nearly 50 percent for large family farms ($250,000-$499,999 annual sales). Operators supply only 19 percent of labor on very large family farms and nonfamily farms.

Tenure

Renting land is a way to expand by controlling additional land without the debt and commitment of capital associated with ownership (Reimund and Gale, 1992, pp. 7-8). About two-thirds of medium-sales and large-scale farms are part owners, meaning that they own part of the land they operate and rent the rest. In addition, 14 percent of large-scale farms—versus 6 percent of all farms—are tenants that own none of the land they farm. About three-quarters of large-scale tenants specialize in crops, compared with two-fifths of farms in general.

Specialization

Specialization varies by farm size. Small farms tend to specialize in raising beef cattle, other grazing livestock, or a variety of field crops (table 2). Poultry, hogs, and high-value crops tend to be produced on large-scale farms. Medium-sales farms and large family farms are most likely to specialize in grain.

Beef cattle

Beef cattle are a common specialization among small farms, accounting for 34 to 41 percent of limited-resource, retirement, residential/lifestyle, and low-sales farms. Beef cattle—commonly cow-calf enterprises in the case of small farms—offer three advantages to operators of small farms. First, cattle are less labor-intensive than many other enterprises, which may be attractive to an operator who is retired or holds a full-time job off the farm (Cash, 2002, p. 21). Second, cattle enterprises tend to be low-cost, which limits cash requirements. Third, under the existing tax code, losses from farming can be written off against income from other sources (Freshwater and Reimer, 1995, p. 220). Producing calves allows farmers to group their expenses and sales in different years to generate small profits in some years and large losses in others (Hoppe and Banker, 2006, p. 14).

Other specializations

Two other specializations were common among limited-resource, retirement, residential/lifestyle, and low-sales farms. About one-quarter of the four groups specialized in "other field crops," which also includes farms with all their crop acres in the CRP and WRP. Another fifth of each group specialize in "other livestock," which includes grazing livestock other than cattle (namely horses, sheep, and goats.)

Some specializations are more common among family farms with gross sales greater than $100,000 (medium-sales and large-scale farms). Farms specializing in cash grains account for about 40 percent of medium-sales and large family farms, while 11-16 percent of medium-sales and large- scale farms specialize in dairy (versus 3 percent of farms in general). Very large family farms are at least twice as likely as any other type to specialize in poultry or hogs, accounting for three-fourths of poultry production and two-thirds of hog production (figure5).

High-value Crops

Production of high-value crops is heavily concentrated among very large family farms and nonfamily farms, which together account for 78 percent of the total. No more than 10 percent of any small farm type specializes in these crops (table 2). High-value crops can generate large sales per acre, but they can require much more labor than cattle and they may require more marketing expertise.

Diversification

Family farms become more diversified as their size increases. Many small family farms specialize in a single commodity or produce nothing at all. Farms with no production include those with all their cropland in the CRP or WRP, as well as farms experiencing crop failure or loss of livestock from disease or other causes. Medium-sales and large-scale farms are more likely to produce multiple commodities: three-fifths of farms in these groups produce three or more commodities.

OPERATOR DEMOGRAPHICS

Every farm has at least one operator, or a farmer who makes day-to-day decisions about the farm business. However, some farms—particularly the larger ones—have more than one operator who makes decisions. In such cases, one operator is designated as the principal operator, the one who is most responsible for running the farm. The others are designated as secondary operators.[4]

Table 2. Farm specialization and diversification, by farm type, 2004

Item	Sm all family farms					Large-scale family farms		Nonfamily farms	All farms
				Farming-occupation					
	Limited-resource	Retirement	Residential/ lifestyle	Low-sales	Medium-sales	Large	Very large		
				Number					
Total farms	197,734	338,671	837,542	395,781	133,299	86,087	71,708	47,103	2,107,925
				Percent					
Commodity specialization:[1]									
Cash grain[2]	11.4	8.2	11.3	14.3	38.7	42.0	24.5	*9.9	14.8
Other field crops[3]	23.2	27.8	23.8	19.6	10.9	11.2	9.9	33.9	22.0
High-value crops[4]	*9.5	5.7	**3.3	9.1	5.9	9.6	12.6	23.4	6.5
Beef	34.3	40.5	37.9	34.0	20.0	11.9	12.1	23.9	33.9
Hogs	d	d	*1.1	*0.6	*3.5	4.2	9.2	1.1	1.6
Dairy	d	d	d	3.0	16.2	13.0	11.3	2.1	2.9
Poultry	d	d	**0.9	d	*2.2	6.5	18.1	#1.2	**1.6
Other livestock[5]	18.5	15.9	*21.3	19.1	#2.6	1.7	*2.2	*4.4	16.7
				Number					
Average number of commodities[6]	1.6	1.4	*1.4	1.9	3.5	3.4	3.2	1.4	1.8
				Percent					
Number of commodities:[6]									
None[7]	14.8	18.6	17.7	13.0	0.0	0.0	d	*23.2	14.4
One	42.4	38.0	40.5	33.5	14.2	13.7	20.4	41.1	35.6
Two	25.4	33.8	28.9	31.4	24.4	22.7	19.1	*22.4	28.8
Three	*9.9	*6.9	#7.8	10.0	17.2	22.4	23.1	*5.9	9.9
Four or more	7.4	#2.7	#5.1	12.1	44.2	41.2	37.4	7.4	*11.4

d = Data suppressed due to insufficient observations.

* = Standard error is between 25 percent and 50 percent of the estimate.

** = Standard error is between 51 percent and 75 percent of the estimate.

= Standard error is greater than 75 percent of the estimate.

[1]Commodity that accounts for at least half of the farm's value of production.

[2]Includes wheat, corn, soybeans, grain sorghum, rice, and general cash grains, where no single cash grain accounts for the majority of production.

[3]Tobacco, peanuts, cotton, sugar beets, sugar cane, corn for silage, sorghum for silage, hay, canola, and general crops, where no single crop accounts for the majority of production. Also includes farms with all cropland in the Conservation Reserve or Wetlands Reserve Programs (CRP or WRP).

[4]Vegetables, fruits and tree nuts, and nursery and greenhouse products.

[5]Includes sheep, lambs, wool, goats, goats' milk, mohair, horses, ponies, mules, donkeys, bees, honey, aquaculture, mink, rabbits, other fur-bearing animals, bison, deer, elk, llamas, etc. Also includes farms where no single livestock species accounts for the majority of production.

[6]Based on 26 commodities or commodity groups: barley, oats, wheat, corn for grain, corn silage, soybeans, sorghum for grain, sorghum silage, canola, fruits, vegetables, nursery products, peanuts, sugar cane, sugar beets, rice, potatoes, cotton, tobacco, hay, other crops, cattle, hogs, dairy, poultry, and other livestock.

[7]Includes farms with no production due to drought, other adverse weather, crop and livestock disease, etc. Also includes farms with all cropland in CRP and WRP.

Source: USDA, Economic Research Service, 2004 Agricultural Resource Management Survey, Phase III.

Principal Operators

Principal farm operators are largely White and male. Minorities account for 5 percent of all principal operators, and a similar percentage for each farm type except for limited-resource farms (table 3). About 12 percent of limited-resource farms had a minority operator, more than double the rate for all farms. Women make up 16 percent of operators on limited-resource farms, nearly double their 9-percent share of all farms.[5] Men operate virtually all family farms with sales of $100,000 or more (medium-sales, large, and very large farms).

Education

Educational attainment varies sharply by type of farm. One-fourth of principal operators of limited-resource farms attended or completed college, compared with half the operators of residential/lifestyle or medium-sales farms. Educational attainment also increases with farm size. About 38 percent of low-sales operators attended or completed college, a number that jumps to 60 percent for operators of very large farms. Thirty-five percent of limited-resource farmers had less than a high-school education, about three times the percentage for all operators.

Operator Age

One of the most striking characteristics of U.S. agriculture is the advanced age of principal farm operators compared with other self-employed workers. About 27 percent of farm operators reported their age as 65 or older in 2004. In contrast, the Bureau of Labor Statistics (BLS) estimates that only 8 percent of self-employed workers in nonagricultural industries were that old (U.S. Dept. of Labor, 2005, p. 220). Each farm type—except residential/lifestyle farms—had a larger share of operators who were at least 65 than was true for the nonfarm self-employed. Retired operators were most likely to be 65 or more, as one would expect, but nearly 60 percent of limited-resource operators were also that old.

The age gap between farm operators and other self-employed workers has increased in recent decades (figure 7). In 1969, 17 percent of farm operators were at least 65 years old, or 6 percentage points more than the estimate for the nonagricultural self-employed. By 2002, the difference had increased to 19 percentage points, largely due to a growing share of older farm operators.

The advanced age of farm operators is understandable, given that the farm is the home for most farmers and that farmers can phase out of farming gradually over a decade or more (Ahearn et al., 1993, p. 7). Younger farmers enter the business at a very slow rate, which tends to increase the average age for farmers as a whole. Improved health and advances in farm equipment have also allowed farmers to farm later in life than in previous generations (Mishra et al., 2005, p. 14).

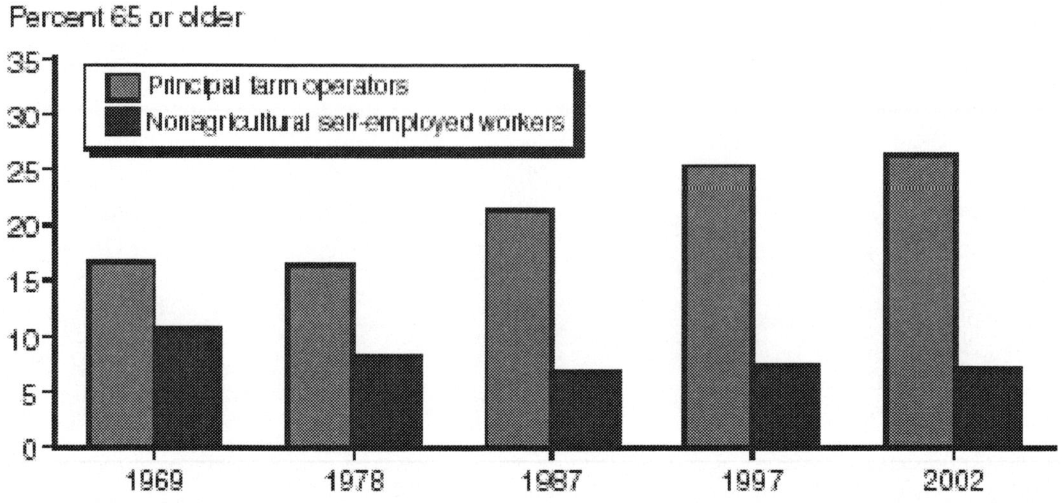

Source: USDA, Economic Research Service, compiled from agricultural census data and from Bureau of Labor Statistics data published in various January issues of *Employment and Earnings*.

Figure 7. Principal farm operators and self-employed workers in nonagricultural industries who were at least 65 years old, selected census years, 1969-2002
Principal farm operators are increasingly likely to be at least 65 years old

The advanced age of farmers raises concerns about a mass exit of farmers from agriculture in the near future (Gale, 2002, p. 30) and finding younger farmers to replace them. Finding replacement operators, however, may not be as hard as it seems (Hoppe and Banker, 2006, p. 36). Older farmers can be replaced with younger farmers producing more on larger farms, and some replacement farmers already work as secondary operators on their families' farms. In addition, about one-fifth of farm operators report they are retired. Any replacement of these operators by younger operators has already occurred.

Secondary Operators and Their Farms

In addition to principal farm operators, there are secondary operators on 967,730 multiple-operator farms (table 4). Because farms are generally family businesses, one would expect family members to serve as secondary operators. In fact, 65 percent of the secondary operators—720,000 out of 1.1 million—are spouses.

The number of operators per farm tends to increase with size. Commercial-sized farms often require more management and labor than an individual can provide. The number of operators per farm reaches 1.9 operators—on average—for very large family farms. Sixty-five percent of farms that size have two or more operators, versus 46 percent for all U.S. farms. About 16 percent of all multiple-operator farms are multiple-generation farms, with at least 20 years' difference between the ages of the oldest and youngest operators.

Table 3. Selected characteristics of principal operators, by farm type, 2004

Item	Small family farms					Large-scale family farms			
				Farming-occupation					
	Limited-resource	Retirement	Residential /lifestyle	Low-sales	Medium-sales	Large	Very large	Nonfamily farms	All farms
	Number								
Total principal operators	197,734	338,671	837,542	395,781	133,299	86,087	71,708	47,103	2,107,925
	Percent of group								
Race or ethnic origin of principal operator:									
White, not Hispanic	88.2	95.5	95.7	95.8	98.0	96.7	96.4	96.1	95.2
Minority[1]	11.8	4.5	4.3	*4.2	*2.0	#3.3	3.6	#3.9	4.8
Gender of principal operator:									
Male	83.6	90.4	91.3	89.0	97.9	98.4	97.5	83.1	90.7
Female	16.4	9.6	**8.7	11.0	2.1	1.6	2.5	*16.9	*9.3
Education of principal operator:									
Some high school or less	35.5	14.0	5.9	11.2	7.1	8.9	5.3	*2.2	11.1
Completed high school	40.6	38.1	39.1	50.6	42.3	38.3	35.2	32.4	41.1
Some college	17.5	24.0	26.5	18.9	29.2	29.2	28.5	21.6	24.0
Completed college	*6.4	24.0	28.6	19.3	*21.3	23.6	31.1	43.7	23.8
	Years								
Average age of principal operator	65	68	51	57	52	52	52	56	56

Table 3. (Continued)

Item	Small family farms					Large-scale family farms			
				Farming-occupation					
	Limited-resource	Retirement	Residential/lifestyle	Low-sales	Medium-sales	Large	Very large	Nonfamily farms	All farms
	Percent of group								
Age of principal operator:									
Younger than 35 years	*2.6	d	6.0	4.1	7.3	6.5	4.9	2.6	4.4
35 to 44 years	*4.7	d	19.3	9.7	18.9	17.9	18.9	15.0	13.0
45 to 54 years	15.4	5.5	37.0	20.5	34.4	35.3	37.6	21.2	26.3
55 to 64 years	19.3	24.8	30.7	40.5	*24.0	26.9	26.5	38.1	29.9
65 years or older	58.3	68.7	7.0	25.3	15.5	13.4	12.1	23.1	26.5

d = Data suppressed due to insufficient observations.

* = Standard error is between 25 percent and 50 percent of the estimate.

**Standard error is between 51 percent and 75 percent of the estimate.

= Standard error is greater than 75 percent of the estimate.

[1]Includes American Indians or Alaska Natives, Asians, Blacks or African Americans, Hispanics or Latins, and Native Hawaiians and other Pacific Islanders. Also includes operators who reported more than one racial or ethnic group. Small sample size for individual minority groups prevents separate estimates for each group.

Source: USDA, Economic Research Service, 2004 Agricultural Resource Management Survey, Phase III.

Table 4. Multiple-operator farms, by farm type, 2004

Item	Limited-resource	Retirement	Residential/ lifestyle	Low-sales	Medium-sales	Large	Very large	Nonfamily Farms	All farms
		Small family farms				Large-scale family farms			
			Farming-occupation						
			Number						
Total operators	273,308	511,094	*1,268,612	590,099	204,683	146,199	136,763	89,290	3,220,048
Principal operators[1]	197,734	338,671	837,542	395,781	133,299	86,087	71,708	47,103	2,107,925
Secondary operators	75,574	172,423	*431,070	194,318	71,384	60,112	65,055	42,187	1,112,123
Spouses	53,754	110,054	*314,752	134,549	45,415	26,804	23,946	*10,542	719,816
Other	*21,820	*62,369	*116,318	59,769	25,968	33,308	41,109	31,645	392,307
Operators (principal and secondary) per farm	1.4	1.5	1.5	1.5	1.5	1.7	1.9	1.9	1.5
Multi-operator farms[2]	71,347	142,740	*392,904	176,010	62,327	48,332	46,841	27,230	967,730
			Percent of group						
Multiple-operator farms as share of all farms	36.1	42.1	46.9	44.5	46.8	56.1	65.3	57.8	45.9
			Number						
Multi-generation farms[3]	*13,524	**33,892	*41,879	19,004	11,855	13,447	12,487	*5,936	*152,025
			Percent of group						
Multiple-generation farms as share of multiple- operator farms	*19.0	*23.7	10.7	10.8	19.0	27.8	26.7	21.8	15.7

Note: ARMS counts all operators—principal and secondary—and asks for detailed information on up to three operators.

* = Standard error is between 25 percent and 50 percent of the estimate.

** = Standard error is between 51 percent and 75 percent of the estimate.

[1]The number of principal operators equals the number of farms. Each farm has one principal operator.

[2]Mulitiple-operator farms report more than one operator.

[3]Farms reporting a difference of at least 20 years between the ages of the youngest and oldest operators.

Source: USDA, Economic Research Service, 2004 Agricultural Resource Management Survey, Phase III

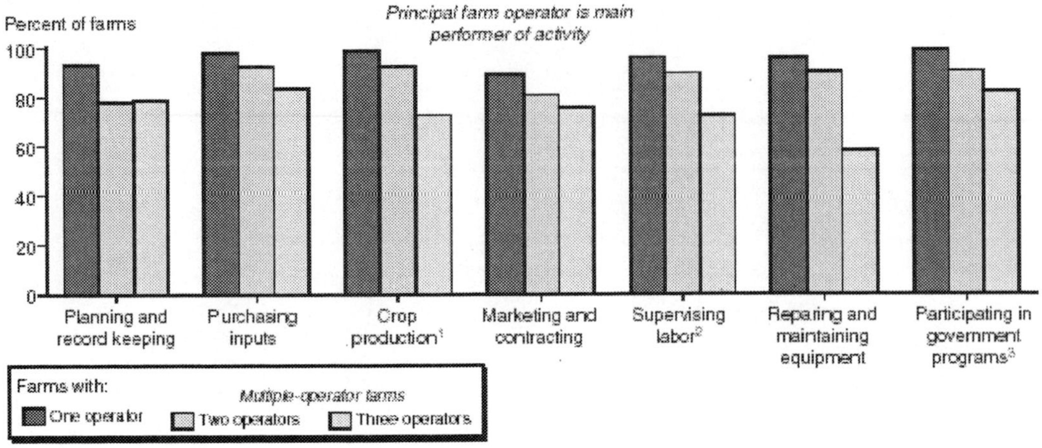

Note: The Agricultural Resource Management Survey collects detailed information on up to three farm operators for each farm.

[1]Estimated only for farms producing crops.

[2]Estimated only for farms hiring labor.

[3]Estimated only for farms receiving government payments.

Source: USDA, Economic Research Service, 2004 Agricultural Resource Management Survey, Phase III, version 1.

Figure 8. Principal farm operators' responsibility for selected activities on large-scale family farms, by number of operators, 2004
Principal operators' responsibilities lessen on multiple-operator farms

The tasks that principal operators perform—like buying inputs, marketing, and maintaining equipment—change as the number of operators increases (figure 8). When there is only one operator on large-scale farms, that operator is the main performer of each activity for 89 to 99 percent of the farms. As the number of operators increases, the percentages fall. For farms with three operators, the principal operator is the main performer for only 58 to 84 percent of the farms, depending on the activity.

As more operators are added, individual operators can specialize in different activities. Or, older operators can scale back and allow younger operators to assume more responsibilities. This appears to be the case on large-scale farms with three operators. The average age of the principal operator on these farms is 59, and 74 percent of the farms are multiple-generation.

FARM INCOME AND FINANCIAL PERFORMANCE

Profitability measures arc strongly associated with farm size. The average operating profit margin and average rates of return on assets and equity are negative for small farms, but positive for large-scale and nonfamily farms (table 5). These ratios are higher for very large farms than for large farms, reflecting very large farms' higher level of sales.

Average profit measures, however, obscure the wide variation in financial performance among farms, including small farms. Although nearly half or more of the farms in each small farm type had a negative operating profit margin in 2004, other small farms were much more profitable (figure 9). For example, between 15 percent and 28 percent of each small farm type had an operating profit margin of at least 20 percent. Nevertheless, an even greater share of large-scale family farms had profit margins that high—36 percent for large family farms and 42 percent for very large family farms. In addition, most of the farms in both of these groups had a positive operating profit margin.

A large majority of each small farm type generated a positive net farm income, although average net farm income was low compared with large and very large family farms (table 5). Overall, net farm income averaged $25,000 per farm in 2004—up 37 percent from the previous year— reflecting a good year for the farm sector. Seventy percent of farms in 2004 earned positive net farm income, and these profitable farms accounted for the bulk of agricultural activity. They generated 81 percent of the total value of production and operated 66 percent of the land in farms.

Selected Financial Ratios

On average, both limited-resource and residential/lifestyle farms had an operating expense ratio greater than 100 percent in 2004. In other words, operating expenses exceeded gross cash farm income. The remaining categories of small farms—retirement, low-sales, and medium-sales—generated enough income to cover expenses. Large-scale family farms and nonfamily farms each had an operating expense ratio of about 70 percent, similar to that of medium-sales farms. Such a ratio provides a more comfortable margin between expenses and income than that experienced by smaller farms.

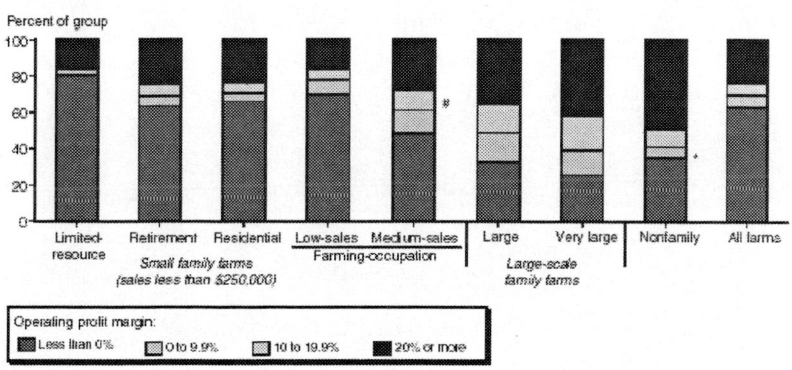

Note: Operating profit margin = 100% X (net farm income + interest paid – charge for unpaid operators' labor and management)/gross farm income.

* = Standard error is between 25 percent and 50 percent of the estimate.

= Standard error is greater than 75 percent of the estimate.

Source: USDA, Economic Research Service, 2004 Agricultural Resource Management Survey, Phase III.

Figure 9. Farms by operating profit margin, 2004

Small family farms are more likely to have a negative operating profit margin

Family farms with annual gross sales of at least $100,000—medium-sales, large, and very large farms—have a higher debt/asset ratio than smaller family farms. As a result, they are also more likely to be marginally solvent (positive net farm income, but with a debt/asset ratio above 40 percent).[6] In contrast, limited-resource, residential, and low-sales small farms are more likely to fall in the marginal-income category (negative net farm income, but with a debt/asset level of no more than 40 percent). This reflects their higher operating expense ratios, which means they are more likely to generate negative net income. Vulnerable farms—with negative net income and a debt/asset ratio above 40 percent—are rare in all farm types, and amount to less than 3 percent of all farms. Residential/lifestyle farms make up 56 percent of the vulnerable group, but their operators are unlikely to depend on the farm for their livelihood.

Most U.S. farms have a favorable financial position, which means they generate positive returns and have a debt/asset ratio no more than 40 percent. Two out of three U.S. farms and at least 60 percent of each farm type were classified as such in 2004.

Loans and Lenders

Many farmers operate with seasonal production loans that are taken out and repaid within the same calendar year, but 41 percent of farms reported outstanding loans as of December 31, 2004 (table 6). This relatively low incidence of debt contributes to the small share of farms with a vulnerable or marginally solvent financial position. Operations with year-end loan balances ranged from one-fifth for retirement farms to three-fourths or more for medium-sales, large, and very large farms. This suggests that the farms most likely to be in debt are larger and most likely to benefit from using credit as a source of capital.

Average debt levels also varied with size, ranging from just over $200,000 for medium-sales farms to nearly $600,000 for very large farms. Smaller family farms averaged less than $100,000 in outstanding debt. Regardless of farm type, real estate and other long-term loans accounted for most debt. Real estate Farms reporting outstanding loans, by lender and farm type, 2004 debt accounted for 84 percent of total debt for residential/lifestyle farms, compared with just over half of total debt on medium-sales and large-scale farms. This high share of real estate debt for residential/lifestyle farms likely reflects substantial mortgages on farm dwellings.

Farm loans originate from a variety of sources. Commercial banks accounted for about 49 percent of the total outstanding loan balances reported by farm operators at the end of 2004. Banks also accounted for a substantial portion of loan balances (37 to 66 percent) for each farm type. The Farm Credit System (FCS) supplied another 25 percent of loan balances..

Table 5. Selected performance measures, by farm type, 2004

	Small family farms					Large-scale family farms			
				Farming-occupation					
Item	Limited-resource	Retirement	Residential/ lifestyle	Low-sales	Medium-sales	Large	Very large	Nonfamily farms	All farms
				Number					
Total farms	197,734	338,671	837,542	395,781	133,299	86,087	71,708	47,103	2,107,925
				Percent					
Profitability measures:									
Rate of return on assets[1]	-4.0	*-1.5	-2.0	- 2.7	#-0.4	2.5	6.8	7.1	**0.5
Rate of return on equity[2]	-4.4	-1.7	-2.8	-3.2	**-1.3	*1.8	6.7	7.1	#-0.1
Operating profit margin[3]	-86.7	*-27.8	-35.5	-36.1	#-2.4	10.8	18.3	23.8	**3.0
				Dollars per farm					
Income measures:									
Net farm income	**1,812	9,655	4,544	9,098	39,084	87,499	287,921	175,795	25,003
				Percent					
Farms with positive net farm income	66.7	79.5	62.8	68.7	76.9	82.2	83.8	72.2	69.6
Financial efficiency measure:									
Operating expense ratio[4]	143.3	83.7	106.8	89.8	74.0	69.1	70.2	70.4	75.3
Solvency measure:									
Debt/asset ratio[5]	*3.9	2.5	3.0	*5.4	10.8	13.1	16.7	7.2	8.8
Solvency and income measure:									
Financial position—[6]									
Favorable	65.9	79.1	60.0	67.0	71.6	71.8	69.2	69.4	66.7
Marginal income	32.3	19.4	33.1	28.9	19.8	15.3	11.3	26.1	27.6
Marginal solvency	d	d	*2.8	*1.7	5.3	10.4	14.6	*2.8	2.9
Vulnerable	d	d	4.0	#2.4	*3.3	2.5	4.9	*1.7	2.8

d = Data suppressed due to insufficient observations.
* = Standard error is between 25 percent and 50 percent of the estimate.
** = Standard error is between 51 percent and 75 percent of the estimate.
# = Standard error is greater than 75 percent of the estimate.

[1]Return on assets = 100% X (net farm income + interest paid - charge for unpaid operators' labor and management) / total assets.
[2]Return on equity = 100% X (net farm income - charge for unpaid operators' labor and management) / net worth.
[3]Operating profit margin = 100% X (net farm income + interest paid - charge for unpaid operators' labor and management) / gross farm income.
[4]Operating expense ratio = 100% X total cash operating expenses / gross cash farm income.
[5]Debt/asset ratio = 100% X total liabilities/total assets.
[6]Financial performance classification based on farm income and debt/asset ratio:

- Favorable: positive net farm income and debt/asset ratio no more than 40 percent.
- Marginal-income: negative net farm income and debt/asset ratio no more than 40 percent
- Marginal-solvency: positive net farm income and debt/asset ratio greater than 40 percent.
- Vulnerable: negative net farm income and debt/asset ratio greater than 40 percent.

Source: USDA, Economic Research Service, 2004 Agricultural Resource Management Survey, Phase III.

Table 6. Farms reporting outstanding loans, by lender and farm type, 2004

Item	Small family farms					Large-scale family farms		Nonfamily farms	All farms
				Farming-occupation					
	Limited-resource	Retirement	Residential/ lifestyle	Low-sales	Medium-sales	Large	Very large		
	Number								
Total farms[1]	#178,433	322,856	824,579	#419,662	138,390	100,870	82,585	40,004	**2,107,377
	Percent								
Share of farms reporting									
outstanding loans[2]	#26.2	17.0	41.4	#41.0	74.4	73.0	82.3	*32.9	**41.4
	Dollars per farm								
For farms with									
outstanding loans:									
Average debt[3]	#50,551	63,154	82,030	#84,908	210,957	*265,223	594,084	*364,334	**153,557
	Percent								
Type of debt:[3]									
Short-term debt	#9.7	**17.9	*5.9	*14.4	21.9	*19.0	26.1	*16.8	*18.0
Real estate	**63.3	68.8	83.6	65.4	55.7	54.1	51.7	60.4	62.0
Other long-term debt	#27.0	*13.3	10.4	*20.2	22.4	*26.9	22.2	*22.7	20.0
Debt-asset ratio[3]	#10.6	11.1	17.5	**12.8	18.7	20.6	**21.2	16.5	*17.8
	Billion dollars								
Outstanding loan									
balances[2]	#2.3	*3.3	30.5	#14.3	19.2	*18.0	39.2	*4.7	*131.6
	Percent of gro up								
Outstanding loans, by lender:[2]									
Farm Credit System	#27.0	**6.6	11.4	**10.3	*27.1	*29.6	37.1	*32.7	*24.6
Commercial banks	#37.0	66.1	57.4	**54.3	*44.7	*48.0	43.2	41.4	*49.0
Farm Service Agency	#2.2	d	*2.4	#6.1	*5.4	*2.0	**2.3	**1.0	*3.1
Life insurance companies	d	d	#0.1	#1.8	#0.1	#0.7	**1.9	#2.4	*1.1
Other[4]	#33.8	22.2	*28.7	#27.5	22.7	*19.6	*15.5	*22.5	#22.2

Table 6. (Continued)

Item	Small family farms					Large-scale family farms			
				Farming-occupation					
	Limited-resource	Retirement	Residential/lifestyle	Low-sales	Medium-sales	Large	Very large	Nonfamily farms	All farms
	Percent of U.S. total								
Outstanding loans, by lender:[2]									
Farm Credit System	1.9	*0.7	10.8	4.6	16.1	16.4	44.7	4.8	100.0
Commercial banks	1.3	3.4	27.2	12.1	13.3	*13.4	26.3	3.0	100.0
Farm Service Agency	1.3	d	17.9	21.5	25.7	8.9	22.5	*1.2	100.0
Life insurance companies	d	d	d	d	d	d	53.1	*7.9	100.0
Other[4]	#2.7	*2.5	*30.0	**13.4	*14.9	*12.1	*20.7	*3.6	100.0
All lenders	#1.8	*2.5	23.2	*10.9	14.6	13.7	29.8	3.6	100.0

d = Data suppressed due to insufficient observations.

* = Standard error is between 25 percent and 50 percent of the estimate.

** = Standard error is between 51 percent and 75 percent of the estimate.

= Standard error is greater than 75 percent of the estimate.

[1]The number of farms differs from that in previous tables because this table is based only on version 1 of the survey.

[2]Based on the outstanding loan balances for up to five loans that were reported on the survey. Information collected about individual loans included the interest rate, purpose of the loan, and lender (identified from a list of 17 potential lenders).

[3]Includes all debt, not just the balances of the five reported outstanding loans. Note that debt/asset ratios in table 6 are higher than those in table 5, because table 6 excludes farms with no outstanding debt.

[4]Loans from the Small Business Administration, State and county government lending agencies, savings and loan associations, implement dealers, financing corporations, input suppliers, cooperatives and other merchants, contractors, other lenders, individuals, and credit cards.

Source: USDA, Economic Research Service, 2004 Agricultural Resource Management Survey, Phase III, version 1.

The Farm Service Agency (FSA) accounts for only 3 percent of all reported balances. Although it makes loans directly to farmers, FSA also guarantees loans made by other lenders. These loans are excluded from estimates of loans held by FSA, but are included in the estimates of debt held by the lenders who made the loan. FSA's direct loans are targeted at beginning farms and farms with smaller credit needs. FSA appears to be serving smaller farms, with 47 percent of its loans going to small farms where the operators report farming as their major occupation. In contrast, only 21 percent of FCS loans and 25 percent of bank loans go to these operations.

Sources and Levels of Operator Household Income

Given their negative operating profit margins and low net farm income—on average—how do so many small farms continue to exist? Households operating small farms typically receive substantial off-farm income. Average off-farm income in 2004 ranged from $13,600 for limited-resource households to $96,900 for households operating residential/lifestyle farms (table 7). Most off-farm income is from earned sources, either a wage-and-salary job or self-employment. However, households operating limited-resource or retirement farms receive well over half their off-farm income from unearned sources—such as Social Security, pensions, dividends, interest, and rent—reflecting the advanced age of operators on those farms. (See "Appendix I: Measuring Operator Household Income and Net Worth" for information on how operator household income is defined.)

Off-Farm Work

Participation in off-farm work varies by farm type. At one extreme, neither the operator nor spouse worked off-farm on 73 percent of limited-resource and 65 percent of retirement farms. At the other extreme, both the operator and a spouse worked off-farm on 64 percent of residential/lifestyle farms. In the remaining farm types, the operator, a spouse, or both worked off-farm in 49 to 62 percent of farm households.

In other words, many farm households are dual-career, holding off-farm jobs as well as farming (Hoppe, 2001, pp. 45 and 49). About 46 percent of all farm households were dual-career, with a spouse working off the farm and the principal operator engaged in farming (with or without off-farm work). According to the Current Population Survey, about 42 percent of all U.S. households had two or more workers in 2004, so farm households are about as likely to be dual-career as U.S. households in general.

Off-farm work is not a recent phenomenon. About 30 percent of principal operators reported off-farm work in the 1930 and 1935 Censuses of Agriculture, generally for fewer than 100 days (figure 10), with considerable variation by State. For example, in the 1935 Census, the share ranged from 18 percent in Iowa and 21 percent in New Jersey to 57 percent in Utah and 60 percent in South Dakota. The percentage working off farm was only 19 percent for South Dakota in the previous (1930) census, with the 5-year jump in off-farm work reflecting "relief work" taken on during drought (Jenkins and Robison, 1937, pp. 8 and 10).

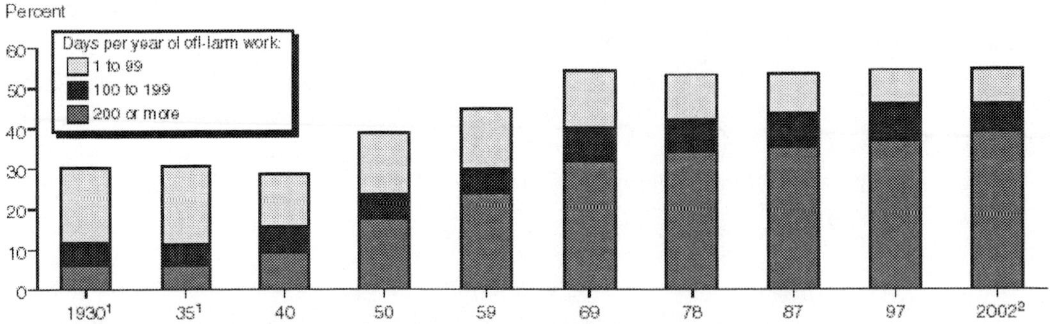

Note: Includes any day the principal operator worked at least 4 hours off the farm.
[1]Excludes Alaska and Hawaii.
[2]Beginning with the 2002 census, days of off-farm work are imputed if not reported.
Source: USDA, Economic Research Service, compiled from census of agriculture data.

Figure 10. Principal farm operators reporting off-farm work, selected census years, 1930 to 2002
Principal farm operators have worked off-farm since at least the 1930s

Since 1969, the share of primary operators working off-farm has stabilized at just over 50 percent nationally. However, the share reporting working 200 days or more—essentially working full-time off the farm—increased steadily from 32 percent in 1969 to 39 percent in 2002.

LEVEL OF PRINCIPAL OPERATOR HOUSEHOLD INCOME

Average principal operator household income was $81,600 in 2004 (table 7), up from $68,500 in 2003, with farming and off-farm income each contributing about half of the $13,100 increase. Households operating large and very large farms experienced substantial increases—$22,700 and $58,300, respectively—mostly from farming. Average farm household income was about 35 percent higher than the average for all U.S. households in 2004, as measured by the Current Population Survey.

Table 7. Income and wealth of principal operator households, by farm type, 2004

	Small family farms					Large- scale family farms		
				Farming-occupation				
Item	Limited-resource	Retirement	Residential/ lifestyle	Low-sales	Medium-sales	Large	Very large	All farms
Number								
Total households	197,734	338,671	837,542	395,781	133,299	86,087	71,708	2,060,822
Dollars per household								
Mean household income	7,680	62,468	96,515	63,043	70,365	125,120	272,527	81,596
Farm earnings[1]	-5,902	*4,128	#-365	**4,925	*34,354	80,250	225,094	14,317
Off-farm income	13,582	58,339	96,879	58,118	36,011	44,870	47,434	67,279
Earned[2]	3,463	20,252	83,548	36,950	26,241	33,238	29,320	48,818
Unearned[2]	10,118	38,087	13,331	21,168	9,769	11,633	*18,114	18,461
Percent								
Share of income from								
off-farm[3]	176.8	93.4	100.4	92.2	51.2	35.9	17.4	82.5
Off-farm work—principal operator and spouse:								
Only operator[4]	18.6	**10.7	35.9	12.3	6.8	8.9	7.3	21.5
Only spouse	*7.2	16.7	0.0	22.5	38.5	33.4	31.8	12.8
Neither[5]	72.6	64.8	0.0	48.0	38.5	45.9	51.1	33.0
Both	d	7.8	64.1	17.2	16.1	11.8	9.8	32.7
Households with:								
Negative farm earnings	72.1	50.8	64.4	44.4	24.6	16.8	16.3	52.8
Negative household income	19.6	*1.2	*0.5	#5.0	13.0	10.5	12.1	5.0
Dollars per household								
Mean household net worth	462,555	685,957	587,960	733,600	1,023,428	1,413,494	2,234,670	739,953
Farm net worth	389,024	464,673	368,200	542,217	910,361	1,237,765	1,975,385	546,788
Nonfarm net worth	73,531	221,284	219,760	191,383	113,067	175,729	259,286	193,165
Percent								
Share of net worth from the farm	84.1	67.7	62.6	73.9	89.0	87.6	88.4	73.9
Real estate share of farm assets[6]	84.8	87.1	83.9	80.0	70.8	68.9	67.8	77.9

Note: Household income and net worth are calculated only for family farms.

d = Data suppressed due to insufficient observations.

*=Standard error is between 25 percent and 50 percent of the estimate.

**Standard error is between 51 percent and 75 percent of the estimate.

= Standard error is greater than 75 percent of the estimate.

[1]Farm earnings in this table and net farm income in table 5 are not directly comparable. Net farm income includes cash and noncash items, is based on accrual accounting, and is calculated for the farm business. Farm earnings—in contrast—are based on cash items only, with the exception of a deduction for depreciation. Farm earnings also exclude the share of net income generated by the farm paid to other households, such as the households of partners. For more information about the definition of farm earnings, see Appendix I.

[2]Earned income comes from off-farm self-employment or wage/salary jobs. Unearned income includes interest and dividends, benefits from Social Security and other public programs, alimony, annuities, net income of estates or trusts, private pensions, regular contributions of persons not living in the household, net rental income from nonfarm properties, and royalties for mineral leases.

[3]Income from off-farm sources can be more than 100 percent of total household income if farm earnings are negative. [4]Includes households were the operator works off-farm and there is no spouse.

[5]Includes households where the operator does not work off-farm and there is no spouse.

[6]Includes farm business assets held by the principal operator's household and other households.

Source: USDA, Economic Research Service, 2004 Agricultural Resource Management Survey, Phase II

Mean income may not be the best choice for comparison, because a few very high income households can raise the mean well above the income received by most households. Nevertheless, using medians still results in higher income for farm households (figure 11). Median farm-operator household income in 2004 was $53,700, 21 percent higher than the median for all U.S. households. Households operating residential/lifestyle, large, or very large farms had a median household income above the median for all U.S. households. The median for retirement and medium-sales households, in contrast, did not differ from the U.S. median by a statistically significant amount. Only two types of farm households—those operating limited-resource or low-sales farms— received median household income below the U.S. median.

Net Worth

The income that farm operator households receive from farming does not reflect the large net worth of many farm households. For example, for households on farms with gross sales of at least $100,000, average net worth ranged from $1 million for medium-sales farms to $2.2 million for very large family farms in 2004 (table 7). Virtually all farm households had a net worth greater than the median net worth for all U.S. households, and nearly two-thirds had a net worth greater than the median for U.S. households with a self-employed head (table 8).

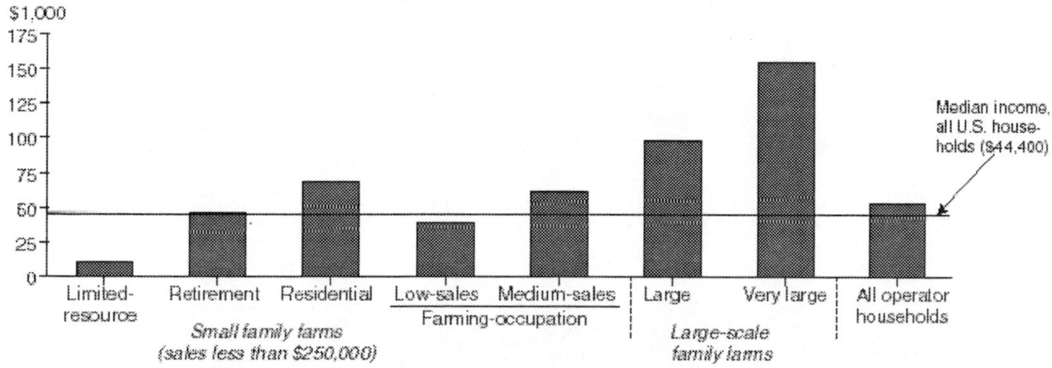

Note: Median income falls at the midpoint of the distribution of households ranked by income. Half of the households have income above the median, while the other half have income below that level.

Source: USDA, Economic Research Service, 2004 Agricultural Resource Management Survey, Phase III, for farm households. U.S. Bureau of the Census, 2005 Current Population Survey for all U.S. households.

Figure 11. Median income of principal-operator households, 2004

Farm households tend to have higher income than U.S. households in general

Table 8. Net worth for farm households versus all U.S. households, 2004

Farm households with a net worth greater than—	All farm households	Limited- resource households
	Percent	
Median for all U.S. households ($93,100)	95.1	85.7
Median for all U.S. households with self-employed head ($335,600)	64.8	42.8

Source: USDA, ERS, 2004 ARMS for farm households. Federal Reserve Board, 2004 Survey of Consumer Finances (Bucks et al., 2006, p. A8) for net worth of all U.S. households and households with a self-employed head.

Even limited-resource households have a relatively high net worth. Eighty-six percent have a net worth higher than the median for all U.S. households, and 43 percent have a net worth greater than the median for households with a self-employed head. The current limited-resource definition has no constraint on farm assets, instead focusing on low household income and low farm sales over a 2-year period. An earlier definition constrained assets, which resulted in a much lower net worth for limited-resource households. For more information, see "Appendix II: Defining Limited-Resource Farms—Past, Present, and Future".[7]

. Unlike income, most of which comes from off-farm sources, net worth from the farm makes up most of the wealth of farm households, regardless of farm type. The farm accounts for 63 to 89 percent of operator household net worth, reflecting the land-intensive nature of farming (table 7). However, much of the net worth of farm households is illiquid—and not available to spend for consumption—because it is largely based on assets necessary to continue farming. Real estate alone amounted to 78 percent of total farm assets.

Government Payments

Farm program payments can be sorted into two groups—commodity-related and conservation (see box, "Types of Farm Program Payments"). Commodity-related payments in total are much larger than conservation payments, accounting for more than four-fifths of all payments made to farmers in the 2004 ARMS data.[8] About 39 percent of farms received government payments of some type in 2004, but the relative shares of government programs varies widely by farm type (figure 12). Medium-sales, large, and very large farms were more likely to receive government payments—especially commodity-related payments—than smaller farms.

Commodity-Related Programs

Commodity programs target specific commodities, largely feed and food grains, cotton, and oilseeds. Payments are tied to the amount of cropland enrolled in programs and yield histories. Specialty crops (except dry peas, lentils, and small chickpeas) and livestock (except dairy, wool, mohair, and honey) are not supported by traditional commodity programs. Producers of nonprogram commodities—as well as producers of program commodities—may also receive disaster assistance and occasional *ad hoc* payments. Farms producing nonprogram commodities may receive substantial payments, if they also produce program commodities or did so in the past.

Most medium-sales and large-scale farms—70 to 80 percent—receive commodity-related payments. These farms collectively received 78 percent of commodity program benefits paid to farmers in 2004, roughly proportional to their share of harvested acres of program crops (figure 13). Very large family farms alone received 35 percent of commodity-related payents.

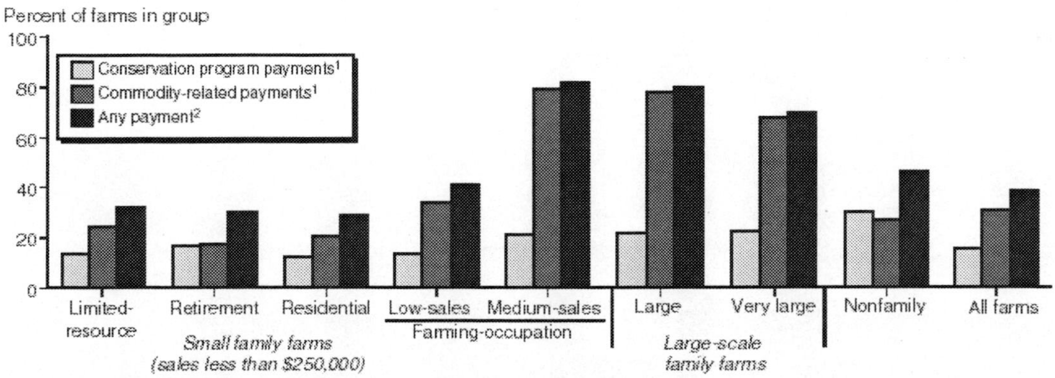

[1]For definitions of conservation program payments and commodity-related payments, see box, "Types of Farm Program Payments."

[2]Receives payments from the conservation programs and/or commodity-related programs. Because some farms receive both types of payments, the percentage of farms receiving commodity-related payments plus the percentage of farms receiving conservation payments sums to more than the percentage of farms receiving any government payment.

Source: USDA, Economic Research Service, 2004 Agricultural Resource Management Survey, Phase III.

Figure 12. Farms receiving payments from conservation or commodity programs, 2004
Most medium-sales and large-scale farms receive payments from commodity programs

Conservation Programs

Nearly 90 percent of conservation payments going to farmers were paid by CRP. The Environmental Quality Incentives Program (EQIP) has expanded since the 2002 Farm Act (Claassen and Ribaudo, 2006), but it still accounted for only 10 percent of conservation payments in the 2004 ARMS. Medium-sales farms received the largest share of EQIP payments, about 46 percent. WRP contributed about 1 percent of conservation payments

Types of Farm Program Payments

The payments covered by the 2004 ARMS are listed below, sorted into two major categories.

Commodity-related

Direct payments, countercyclical payments, loan deficiency payments, marketing loan gains, net value of commodity certificates, peanut quota buyout, milk income loss contract payments, agricultural disaster payments, and any other State, Federal, and local payments.

Conservation

Payments from the Conservation Reserve Program (CRP), Wetlands Reserve Program (WRP), and Environmental Quality Incentives Program (EQIP).

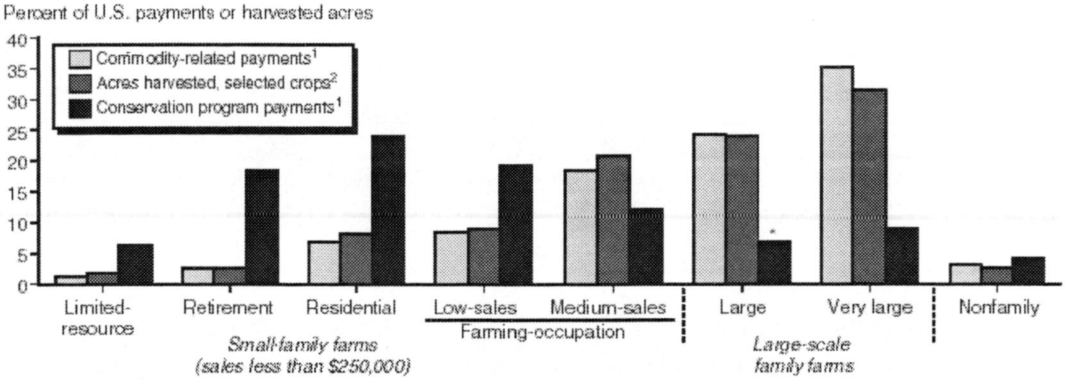

* = Standard error is between 25 percent and 50 percent of the estimate.

[1]For definitions of commodity-related payments and conservation program payments, see box "Types of Farm Program Payments."

[2]Food and feed grains, soybeans, other oilseeds, cotton, and peanuts.

Source: USDA, Economic Research Service, 2004 Agricultural Resource Management Survey, Phase III.

Figure 13. Distribution of payments from conservation and commodity programs, 2004
Acres of program commodities explain the distribution of commodity program payments

The target of CRP (and WRP) is environmentally sensitive land, rather than the production of specific commodities, so the distribution of conservation payments differs from that of commodity-related payments. Retirement, residential/lifestyle, and low-sales farms received 62 percent of conservation payments in 2004, reflecting their large numbers (75 percent of all farms), their large share of farmland (43 percent of the land owned by farms), and their tendency to enroll large shares of their land in CRP and WRP when they do participate. CRP and WRP enrollments account for 47 percent of the land operated on participating retirement farms, 35 percent on participating residential/lifestyle farms, and 25 percent on participating low-sales farms. In contrast, enrollment ranges from 6 percent to 11 percent for participating medium-sales farms and large-scale farms.

Residential/lifestyle operators' main reported occupation is off the farm, which limits the amount of time they spend farming. Since land enrolled in CRP and WRP requires little labor or capital investment and provides a guaranteed income stream, residential/lifestyle farmers may find the programs financially attractive, particularly if their farms are not highly profitable. Given their age, many retired farmers and older low-sales farmers have land available to put into conservation uses (Lambert et al., 2006, pp. 20-26).

SPECIAL FEATURE: THE SHIFT TO LARGE FARMS

During the past two decades, data from the census of agriculture show a marked increase in the number of farms selling at least $250,000 in farm products. The growth in the number of these large farms was accompanied by a similar shift in production. We sort farms in each of the five censuses of agriculture—1982, 1987, 1992, 1997, and 2002—into sales classes in order to track these changes.

When using agricultural sales to measure trends in farm size over time, it is important to adjust for changes in agricultural prices, which will change revenue without any changes in the physical volume of production. Accordingly, we adjust sales of agricultural products for price changes using the Producer Price Index (PPI) for farm products, which is also the USDA/NASS index of prices received by farmers. Sales classes from the various censuses of agriculture presented here are expressed in 2002 constant dollars.

Distribution of Farms

The number of farms with sales of at least $250,000 grew steadily from 1982 to 2002 (table 9), increasing from 85,000 to 152,000. The share of all farms in this group grew from 4 percent to 7 percent. Most of these farms had sales between $250,000 and $499,999—even by the end of the period— but the number of farms with sales of at least $500,000 grew more rapidly. The number of farms with sales between $500,000 and $999,999 more than doubled, while the number of million-dollar farms more than tripled.

The number of farms in the other sales classes declined in each of the four intercensus periods, with the exception of farms selling less than $10,000. Farms with sales that low declined during the first two intercensus periods, but increased during the last two periods. The increase during the last two periods was due to growth in "point farms," or farms with sales less than $1,000 that might normally have sales that high and satisfy the criteria necessary to be considered a farm.[9] Because of the growth in point farms, farms with sales less than $10,000 now account for 59 percent of all U.S. farms, up from 49 percent in 1982.

Most of the increase in point farms, however, was due to a minor change in the census farm definition and an adjustment for undercoverage in the census farm count. Beginning in 1997, establishments that enrolled all their cropland in CRP or WRP were counted as farms, even if they did not sell at least $1,000 in farm products (Hoppe and Korb, 2002, p. 25).[10] With the 2002 census, NASS adjusted the census farm count to compensate for undercoverage (Allen, 2004), which had the largest effect on farms near the $1,000 cutoff in the farm definition (USDA, NASS, 2004, p. C-1 1).[11]

Distribution of Agricultural Sales

In addition to the shift in the number of farms in the various sales classes, even more dramatic shifts occurred in the distribution of total agricultural sales. The share of total sales accounted for by farms with sales of $250,000 or more increased steadily from 47 percent in 1982 to 76 percent in 2002 (figure 14). Farms with sales of $1,000,000 to $4,999,999 and $5 million or more doubled their share of sales between 1982 and 2002. The two largest sales classes now account for nearly one-fourth of agricultural sales each, although the two groups together make up only 1 percent of farms.

Farms with sales of at least $5 million specialized in relatively few commodities in 2002. About 34 percent specialized in high-value crops, with cattle feedlots (19 percent), dairy (14 percent), and poultry/eggs (14 percent) also common. Farms with sales between $1,000,000 and $4,999,999 tended to specialize in a wider variety of commodities: high- value crops (26 percent), poultry and eggs (19 percent), dairy (13 percent), hogs and pigs (11 percent), grains and oilseeds (9 percent), and field crops other than grain (8 percent).

Larger shares of the two sales classes were located in the Pacific region than in any other region: 22 percent for farms with sales between $1,000,000 and $4,999,999 and 32 percent for farms with sales of $5 million or more (table 10). California alone had 17 percent of the farms in the former sales class and 26 percent of the $5-million-plus farms. About 60 percent of California farms with sales of $1 million or more specialized in high-value crops, and another 24 percent specialized in dairy.

Typical Enterprise Size

The shift of sales to larger sales classes is also reflected by an increase in "typical enterprise size" over time. The typical enterprise size aims to capture the size of farm enterprise from which most of a particular commodity came. Specifically, we define it as the median (midpoint) of the distribution of production by enterprise size. For crops, the median defined here identifies the enterprise size at which half of a commodity's harvested acreage came from larger enterprises and half came from smaller enterprises. For example, the typical enterprise size for corn in 2002 of 450 acres (table 11) means that half of all harvested acres of corn is on farms harvesting more than 450 acres of corn and half is on farms harvesting less than 450 acres.[12]

Table 9. Number of farms by constant-dollar sales class (2002 dollars), 1982 to 2002

Constant-dollar sales class (2002 dollars)[1]	Census year					Intercensus period			
	1982	1987	1992	1997	2002	1982 to 1987	1987 to 1992	1992 to 1997	1997 to 2002
	Number of farms					*Percent change*			
Total farms	2,240,976	2,087,759	1,925,300	1,911,859	2,128,982	-6.8	-7.8	-0.7	11.4
Less than $10,000	1,106,092	1,016,863	927,234	1,009,084	1,263,052	-8.1	-8.8	8.8	25.2
Point farms[2]	254,097	235,562	212,580	277,248	570,919	-7.3	-9.8	30.4	105.9
Other farms	851,995	781,301	714,654	731,836	692,133	-8.3	-8.5	2.4	-5.4
$10,000 to $49,999	586,007	547,150	490,530	430,065	414,063	-6.6	-10.3	-12.3	-3.7
$10,000 to $19,999	257,391	251,361	228,504	204,384	197,967	-2.3	-9.1	-10.6	-3.1
$20,000 to $24,999	79,954	76,069	68,069	58,444	58,190	-4.9	-10.5	-14.1	-0.4
$25,000 to $39,999	167,510	149,905	133,059	115,582	109,310	-10.5	-11.2	-13.1	-5.4
$40,000 to $49,999	81,152	69,815	60,898	51,655	48,596	-14.0	-12.8	-15.2	-5.9
$50,000 to $99,999	250,694	217,871	187,062	157,635	140,479	-13.1	-14.1	-15.7	-10.9
$100,000 to $249,999	213,264	207,999	202,779	179,091	159,052	-2.5	-2.5	-11.7	-11.2
$250,000 or more	84,919	97,876	117,695	135,984	152,336	15.3	20.2	15.5	12.0
$250,000 to $499,999	57,691	64,195	74,354	78,330	81,694	11.3	15.8	5.3	4.3
$500,000 to $999,999	18,242	22,058	28,583	36,469	41,969	20.9	29.6	27.6	15.1
$1,000,000 to $2,499,999	6,494	8,409	10,634	15,448	20,724	29.5	26.5	45.3	34.2
$2,500,000 to $4,999,999	1,448	1,811	2,392	3,386	4,611	25.1	32.1	41.6	36.2
$5,000,000 or more	1,044	1,403	1,732	2,351	3,338	34.4	23.4	35.7	42.0

Note: Constant-dollar sales classes cannot be prepared before 1982 due to incomplete census records for individual farms prior to that year.

[1]Sales class is expressed in constant 2002 dollars, using the Producer Price Index for farm products to adjust for price changes.

[2]Point farms have sales of less than $1,000 (current dollars), but are still considered farms because they would be expected to normally sell at least $1,000 of agricultural products. Point farms are defined here in current dollars—rather than constant dollars—because they are identified in each census based on current dollars.

Source: USDA, Economic Research Service, compiled from census of agriculture data.

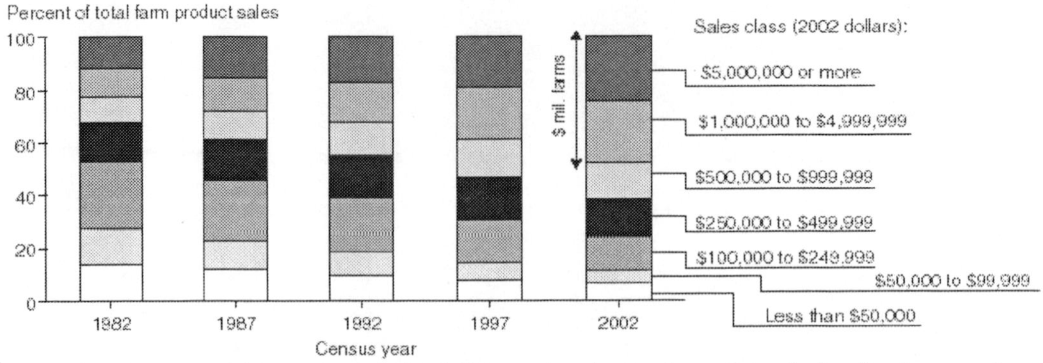

[1]Sales class is expressed in constant 2002 dollars, using the Producer Price Index for farm products to adjust for price changes.

Source: USDA, Economic Research Service, compiled from census of agriculture data.

Figure 14. Distribution of farm product sales by constant-dollar sales class[1] (2002 dollars), 1982-2002
Million-dollar farms' share of sales increased from 23 percent in 1982 to 48 percent in 2002

Table 10. Farms with sales of at least $1 million, by region, 2002

Item	Sales of $1,000,000 or more		
	Total	**$1,000,000 to $4,999,999**	**$5,000,000 or more**
	Number		
Farms	28,673	25,335	3,338
	Percent of U.S. total		
Farms by region:			
Northeast	6.2	6.3	5.1
Lake States	8.3	8.8	4.6
Corn Belt	12.9	13.6	7.5
Northern Plains	8.4	8.0	12.1
Appalachian	9.0	9.5	5.0
Southeast	11.1	11.2	10.2
Delta	6.3	6.9	1.9
Southern Plains	6.5	6.2	8.5
Mountain	8.2	7.6	13.0
Pacific	23.0	21.8	32.0

Northeast: CT, DE, ME, MD, MA, NH, NJ, NY, PA, RI, and VT; Lake States: MI, MN, and WI; Corn Belt: IL, IN, IA, MO, and OH; Northern Plains: KS, NE, ND, and SD; Appalachian: KY, NC, TN, VA, and WV; Southeast: AL, FL, GA, and SC; Delta: AR, LA, and MS; Southern Plains: OK and TX; Mountain: AZ, CO, ID, MT, NV, NM, UT, and WY; Pacific: AK, CA, HI, OR, and WA.

Source: USDA, Economic Research Service, compiled from the 2002 Census of Agriculture.

For dairy, the measure captures the midpoint of the distribution of cows by herd size—half of dairy cows are in larger operations and half are in smaller operations. For poultry and other livestock, the measure captures the midpoint of broiler, cattle, or hog sales by enterprise size. Enterprise size differs from farm size, because a farm may have multiple crop or livestock enterprises.

The well-documented shift to larger livestock enterprises is most evident for hogs. The typical enterprise size increased nearly twentyfold, from sales of 1,200 head in 1987 to 23,400 in 2004. This increase is consistent with the rapid consolidation of the hog industry occurring in recent years (McBride and Key, 2003, pp. 5-10). Typical fattened cattle and dairy enterprises also grew in size, approximately doubling and tripling (respectively) during the 15-year period.

The growth in size between 1987 and 2002 was less extreme for broiler and cow/calf enterprises, around 70 percent for both enterprises. Much of the growth in broiler enterprise size occurred long before 1987, and cow-calf enterprises are a common specialization for small farms. The typical size for cow-calf enterprises is still just 84 calves per year.

Enterprise size has also increased in crop production. Typical acres harvested roughly doubled for each of the field crops, for most types of vegetables, and for oranges. Peach enterprises have been more stable, increasing by only 14 percent between 1987 and 2002.

Business Organization

Despite the shift in farm product sales to larger farms and increasing enterprise sizes, most farms continue to be organized as sole proprietorships, partnerships, or family corporations. These farms have consistently made up about 99 percent of the farm count since 1978 (figure 15), the initial year of the current census series on business organization. They also accounted for more than 90 percent of agricultural sales each year. Marked shifts have occurred in the distribution of sales among these farms between 1978 and 2002, however. Family corporations' share of sales grew by 7 percentage points, and partnerships' share grew by 2 percentage points, while proprietorships' share shrank by 10 percentage points. Nevertheless, sole proprietorships still accounted for 90 percent of farms and 52 percent of sales in 2002.

Nonfamily corporations make up a relatively minor and stable share of farm numbers and sales. Nonfamily corporations—part of the "other organization" category in figure 15—accounted for 0.2-0.4 percent of all farms and 6-7 percent of agricultural sales each census year. Most of these nonfamily corporations are not large, publicly held companies. Between 80 and 87 percent of them, depending on the year, had no more than 10 stockholders.

Regardless of farm type, proprietorships make up the bulk of family farms: approximately 90 percent of each small farm type, 77 percent of large farms, and 60 percent of very large farms (table 12). Given the age distribution of farmers, one would expect to find some farms in estates or trusts. In fact, 50 percent of nonfamily farms fall in the "other organization" category, which includes estates and trusts.

Table 11. Typical enterprise size for selected commodities, 1987 to 2002

	Census year				Change, 1987 to 2002
Selected commodity	1987	1992	1997	2002	
	Typical annual sales[1] (head per farm)			*Percent*	
Poultry/livestock:					
Broilers	300,000	384,000	480,000	520,000	73.3
Hogs	1,200	1,880	11,000	23,400	1,850.0
Fattened cattle	17,532	23,891	38,000	34,494	96.7
Cattle, less than 500 pounds	50	60	65	84	68.0
	Typical herd size[2] (head per farm)			*Percent*	
Dairy production	80	100	140	275	243.8
	Typical acres harve sted[3] (acres per farm)			*Percent*	
Field crops:					
Corn	200	300	350	450	125.0
Soybeans	243	300	380	480	97.5
Wheat	404	562	693	784	94.1
Cotton	450	605	800	920	104.4
Rice	295	400	494	607	105.8
Vegetables:					
Asparagus	160	200	200	236	47.5
Lettuce	949	1,168	1,461	2,225	134.5
Bell peppers	88	130	180	200	127.3
Potatoes	350	422	556	810	131.4
Sweet corn	100	120	173	222	122.0
Tomatoes	400	450	589	700	75.0
Tree crops:					
Apples	83	94	122	129	55.4
Almonds	203	234	292	361	77.8
Oranges	450	732	769	1,015	125.6
Peaches	92	95	100	105	14.1

Note: Census records do not have all the data necessary to derive typical enterprise size prior to 1987.

[1]Median head sold. Half of the sales of a given species were from farms with more than the typical sales and half were from farms with less than the typical sales.

[2]Median head of dairy cows as of December 31 of the census year. Includes dry cows and cows in milk. Half of the cows were on farms with more than the typical number of cows and half were on farms with less than the typical number of cows.

[3]Median acres harvested. Half of all harvested acres of a commodity were on farms harvesting more than the typical number of acres and half were on farms harvesting less than the typical number of acres.

Source: USDA, Economic Research Service, compiled from census of agriculture data.

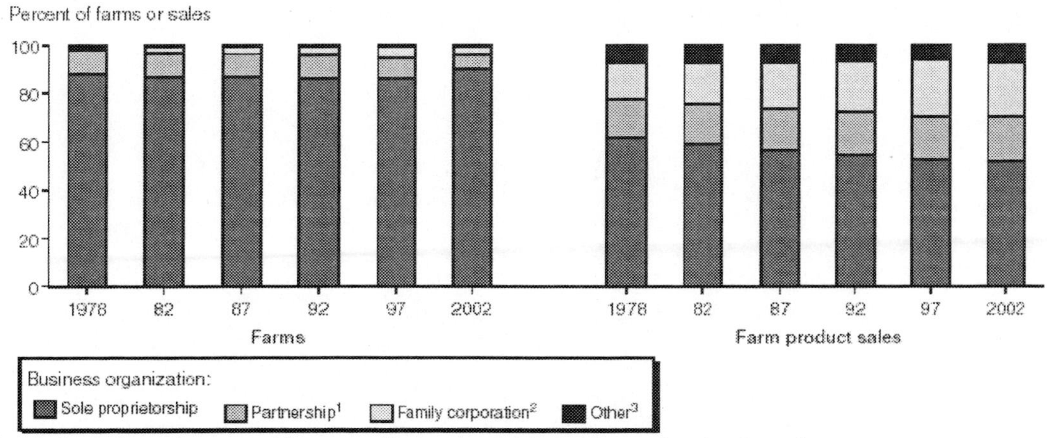

[1]Includes informal partnerships as well as partnerships registered under State law.

[2]Prior to the 2002 census, family-held corporations were defined in the questionnaire as having more than 50 percent of their stock owned by persons related by blood or marriage. No specific definition was used in the 2002 census.

[3]Includes nonfamily corporations, cooperatives, estates or trusts, institutional farms, etc.

Source: USDA, Economic Research Service, compiled from census of agriculture data.

Figure 15. Distribution of farms and farm product sales, by business organization, 1978-2002
Family corporations' share of sales grew the most

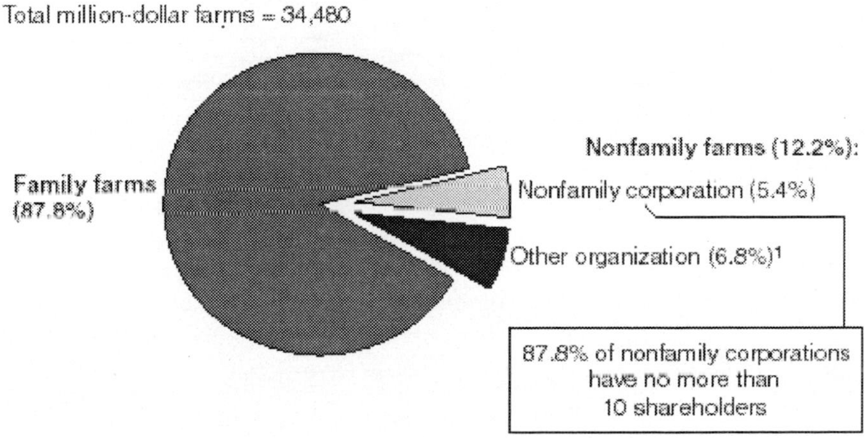

[1]Proprietorships, partnerships, or family corporations with hired managers. Also includes estates, trusts, and cooperatives.

Source: USDA, Economic Research Service, 2004 Agricultural Resource Management Survey, Phase III. (Number of shareholders is from version 1 of ARMS.

Figure 16. Organization of farms with gross sales of $1 million or more, 2004
Most million-dollar farms are organized as family farms

Only 19 percent of nonfamily farms are organized as nonfamily corporations. Direct ownership of large farms by large, publicly held corporations is negligible and is likely to remain so. For example only 5 percent of the 34,500 million-dollar farms were organized as nonfamily corporations in 2004, and 88 percent of these corporations had no more than 10 shareholders (figure 16).

CONTRACTING

Although few nonfamily corporations—large or small—directly operate farms, they often make contracts with farmers to provide the commodities they need for processing or wholesaling. ERS identifies two types of contracts in ARMS:

Production contract. A production contract is a legal agreement between a farm operator (contractee) and another person or firm (contractor) to produce a specific type, quantity, and quality of agricultural commodity. The contractor usually owns the commodity being produced and the farm receives a service fee.

Marketing contract. Under a marketing contract, the contractor buys a known quantity and quality of a commodity from a farm for a negotiated price. The farm owns the commodity while it is being produced and receives a price reflecting the value of the commodity.

Contracts can provide benefits to both producers and contractors (MacDonald and Banker, 2005, pp. 52-53; MacDonald et al., 2004, pp. 24- 30). Farmers get a guaranteed outlet for their production with known compensation, while contractors get an assured supply of commodities with specified characteristics, delivered in a timely manner.

PRODUCTION UNDER CONTRACT

Although production and marketing contracts account for about two-fifths of U.S. agricultural production, the share varies by commodity (figure 17). For example, U.S. farmers produce virtually all sugarbeets and poultry under contract. Contracting also accounts for at least half of the production of cotton, tobacco, fruits, dairy products, and hogs. At the other extreme, only a small portion of wheat, soybeans, or corn—all traditional field crops—is grown under contract.

The aggregate data show slow and steady growth in contracting over the years, but change can be more rapid for some commodities. For example, the share of tobacco production covered by contracts went from 1 percent to 50 percent between 1995-96 and 2004. Cigarette manufacturers replaced cash auctions with contract marketing because contracts better enabled them to acquire enough of the specific types of tobacco they needed. The contracting share of hogs also increased rapidly over this 10-year span, from 31 percent to 71 percent, driven in part by product differentiation. Processors wanted more control over the characteristics of the hogs they acquired, which helped them provide a consistent quality of meat to consumers (MacDonald and Banker, 2005, pp. 55-59).

Variation by Type of Farm

Use of contracts also varies by farm type. The share of limited-resource, retirement, and residential/lifestyle farms using contracts is just 3 or 4 percent (table 13). For the remaining types of family farms, the use of contracts increases with sales, ranging from 9 percent of low-sales farms to 64 percent for very large family farms. The share of their production under contract also increases with sales, from 18 to 51 percent.

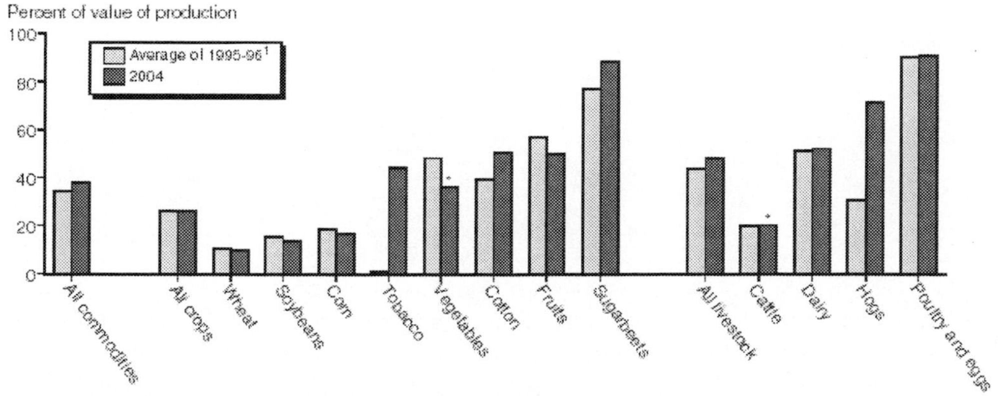

*=Standard error is between 25 percent and 50 percent of the estimate.

[1]An average of 1995 and 1996 was used to provide a more statistically reliable estimate.

Source: USDA, Economic Research Service, 1995 Farm Costs and Returns Survey and 1996 and 2004 Agricultural Resource Management Survey, Phase III.

Figure 17. Share of value of production under marketing or production contracts for selected commodities, 1995-96 and 2004
Share of tobacco and hogs sold or removed under contract increased dramatically

Although a small percentage of each small farm type has contracts, small farms make up 58 percent of the farms with contracts, reflecting their large numbers. Value of production under contract, in contrast, is concentrated among very large family farms, which account for 61 percent of the total.

The value of commodities removed under production contracts is counted in the farms' gross sales, often used as a basic measure of farm size. But the value of commodities removed is not included in gross cash income (equivalent to gross revenue) received by the farms, because they never owned or sold the commodities. Only the fees that the farms receive under a production contract are included in gross cash income. As a result, gross sales are much higher than gross cash income for farms with most of their output under production contracts, such as poultry farms. If gross cash income were used to measure farm size, only 11 percent of poultry farms would be classified as large-scale— using a $250,000 cutoff—compared with 56 percent if gross sales were used (see box, "Gross Sales or Gross Cash Income?").

Table 12. Business organization of farms, by farm type, 2004

Item	Small family farms					Large-scale family farms			
				Farming-occupation					
	Limited-resource	Retirement	Residential/lifestyle	Low-sales	Medium-sales	Large	Very large	Nonfamily farms	All farms
				Number					
Total farms	197,734	338,671	837,542	395,781	133,299	86,087	71,708	47,103	2,107,925
				Percent					
Farms by organization:									
Sole proprietorship[1]	96.1	94.7	92.8	93.1	87.3	77.3	59.8	14.6	89.6
Partnership[2]	1.8	1.7	4.6	4.1	6.8	11.3	18.3	*6.0	4.7
Corporation	d	#3.7	*2.6	2.7	*5.9	11.4	21.9	29.3	*4.6
Family[3]	d	#3.7	*2.6	2.7	*5.9	11.4	21.9	10.0	*4.1
Nonfamily[3]	na	na	na	na	na	na	na	19.3	0.4
Other organization[4]	na	na	na	na	na	na	na	50.1	1.1
Farm product sales									
by organization:									
Sole proprietorship[1]	93.5	87.4	89.8	90.2	86.8	76.8	51.5	**15.1	57.7
Partnership[2]	*2.7	*3.7	6.4	*6.1	*6.7	11.5	20.8	*6.4	13.7
Corporation	d	*8.9	**3.8	3.7	*6.4	11.7	27.7	58.6	25.4
Family[3]	d	*8.9	**3.8	3.7	*6.4	11.7	27.7	14.7	18.2
Nonfamily[3]	na	na	na	na	na	na	na	44.0	7.2
Other organization[4]	na	na	na	na	na	na	na	*19.9	*3.2

d = Data suppressed due to insufficient observations.

na = Not applicable.

* = Standard error is between 25 percent and 50 percent of the estimate.

** = Standard error is between 51 percent and 75 percent of the estimate.

= Standard error is greater than 75 percent of the estimate.

[1]Includes informal partnerships, such as those between spouses. (In the census of agriculture, informal partnerships are classified as partnerships.)

[2]Includes only partnerships registered under State law.

3A corporation is classified as a family corporation if more than 50 percent of the stock is held by people related by blood or marriage. Other corporations are classified as nonfamily.

[4]Estates, trusts, and cooperatives.

Source: USDA, Economic Research Service, 2004 Agricultural Resource Management Survey, Phase III.

Table 13. Farms with contracts and value of production under contract, by farm type, 2004

		Small family farms				Large-scale family farms			
				Farming-occupation					
Item	Limited-resource	Retirement	Residential/ lifestyle	Low-sales	Medium-sales	Large	Very large	Non family farms	All farms
	Number								
Total farms	197,734	338,671	837,542	395,781	133,299	86,087	71,708	47,103	2,107,925
	Percent of group								
Farms with contracts[1]	*2.7	3.3	4.2	9.0	34.5	50.3	63.7	15.4	10.9
Value of production under contract[2]	*10.1	13.3	10.4	*18.2	21.4	34.5	51.0	35.1	37.8
	Percent of U.S. total								
Farms with contracts[1]	2.3	4.9	15.3	15.5	20.0	18.9	19.9	3.2	100.0
Value of production under contract[2]	0.3	0.7	1.5	2.7	6.1	13.5	61.2	14.1	100.0

* = Standard error is between 25 and 50 percent of the estimate.

[1]Includes farms with production contracts, marketing contracts, or both.

[2]Includes commodities under production or marketing contracts.

Source: USDA, Economic Research Service, 2004 Agricultural Resource Management Survey, Phase III.

CONCLUSIONS

This report has four major findings important to understanding farms and farm households today and in the future.

Farm product sales have shifted to larger farms over the past two decades. Farms with sales of $250,000 or more accounted for 76 percent of all sales in the 2002 Census of Agriculture, and million-dollar farms alone accounted for 48 percent.

Most U.S. farms—including million-dollar farms—are family farms. The share of farm output from large, publicly held corporations remains minimal.

GROSS FARM SALES OR GROSS CASH INCOME

Gross farm sales (or gross sales) is an indicator of farm size. It measures what the farm produces, regardless of who has a claim on that production. Gross sales is calculated as the farm's crop and livestock sales plus the shares of production received by any share landlords and production contractors. The measure also includes all government payments received by the farm and its landlords.

In contrast, ***gross cash farm income*** (or gross cash income) is the total revenue received by the farm business alone, excluding any shares accruing to share landlords and contractors. Gross cash income is the sum of livestock sales, crop sales, government payments, and "other farm-related income" received by the farm business. Other farm-related income includes income from a variety of sources: custom work, machine hire, livestock grazing, timber sales, outdoor recreation, contract production fees, etc.

For farms with no production contracts and no landlords, gross sales and gross cash income will generally be the same, both calculated as the sum of crop sales, livestock sales, and government payments received by the farm. In some cases, however, gross cash farm income is higher than gross sales, due to the additional miscellaneous items making up other farm-related income.

For farms with production contracts, gross cash income may be substantially less than gross sales. Commodities removed under production contracts are excluded from gross cash income but are included in gross sales. Fees received from contractors are included in gross cash income—as part of other farm-related income—but these fees are small compared with the value of the commodities removed.

Farms specializing in poultry or hogs have especially small gross cash income, relative to gross sales (see text table below). The ratio of gross cash income to gross sales is lower for poultry farms (34 percent) than for hog farms (72 percent) because poultry farms produce more under product contracts.

Our perception of the size of poultry farms would change if we measured size by gross cash income instead of gross sales. Only 11 percent of poultry farms would be considered large-scale—applying the $250,000 cutoff to gross cash income—instead of 56 percent.

For poultry farms, gross cash farm income was only one-third of gross sales in 2004

Item	Poultry farms	Hog farms	Other farms	All farms
		Number		
Total farms	**34,149	33,292	2,040,485	2,107,925
		Dollars per farm		
Gross farm sales	*685,750	435,882	88,342	103,509
Gross cash farm income	*231,239	314,701	93,574	99,297
		Percent		
Ratio of gross cash farm income to gross farm sales	33.7	72.2	105.9	95.9
Share of production under production contract	85.7	58.9	5.5	18.2
Farms with gross farm sales of $250,000 or more	*55.5	32.0	6.8	8.0
Farms with gross cash farm income of $250,000 or more	*11.3	21.4	7.4	7.7

* = Standard error is between 25 and 50 percent of the estimate.
** = Standard error is between 51 and 75 percent of the estimate.
Source: USDA, ERS, 2004 ARMS.oss Sales or Gross Cash Income?

Generally, large and very large family farms are viable economic businesses, with favorable financial ratios. Small farm businesses are less viable as businesses, but the households operating them receive substantial off-farm income.

Different farm policies affect different sets of farmers. Payments from commodity programs tend to flow to medium-sales and large-scale farms, and conservation payments tend to flow to smaller family farms. A majority of farms, however, receive no government payments, but they may be indirectly affected by the effects of government payments on farmland and commodity markets.

Shifts to Larger Farms

Constant-dollar sales class data show a steady growth in large farms (sales at least $250,000) and decline in the number of farms in most other sales classes. Growth in the number of large farms was accompanied by a sales shift in the same direction. The share of production accounted for by large farms grew from 47 percent in 1982 to 76 percent in 2002.

By 2002, million-dollar farms alone accounted for 48 percent of sales, compared with 23 percent in 1982.

The only other increase in farm numbers was for farms with sales less than $10,000, which grew in the last two intercensus periods because of growth in the number of point farms. Most of this increase was due to a change in the census farm definition and an adjustment for undercoverage in the census farm count. Farms with sales less than $10,000 accounted for 59 percent of farms in 2002, up from 49 percent in 1982. Their share of sales, however, declined from 3 percent in 1982 to 1 percent in 2002. Thus, the 29-percentage-point increase in the share of sales for large farms came largely from a declining share for farms in the $10,000-$249,999 sales classes.

The Place of Family Farms in U.S. Agriculture

Family farms dominate U.S. agriculture. Most farms (98 percent) are family farms, and they collectively generate 85 percent of the value of production. Large-scale family farms account for about 60 percent of production, which is large compared with their 8-percent share of farms. Small family farms make significant contributions to the production of specific commodities, such as wheat, corn, soybeans, hay, tobacco, and beef.

Even million-dollar farms are overwhelmingly family operations. About 88 percent operated as family farms in 2004, and only 5 percent were organized as nonfamily corporations, usually with no more than 10 stockholders. Direct ownership of million-dollar farms by large, publicly held corporations is negligible and is likely to remain that way, although these corporations often act as contractors. Nonfamily corporations made up less than 1 percent of farms and no more than 7 percent of sales in the last six agricultural censuses, despite the ongoing shift of production to large farms.

Financial Status of the Family Farm

Farming had a very good year in 2004. Total net farm income for the sector was $83 billion, substantially higher than the annual average for the previous 10 years ($55 billion) and the previous peak in 1996 ($69 billion), all measured in constant 2004 dollars. Seventy percent of all farms in 2004 earned a positive net farm income. Only 3 percent of farms were classified as vulnerable (negative net cash farm income with a debt/asset ratio greater than 40 percent). More than half of the vulnerable farms were residential/lifestyle farms, whose operators—by definition—rely on off-farm work for their livelihood.

For the most part, large and very large family farms are viable economic businesses. Their average profit margin and rates of return on assets and equity were all positive, and the large majority of these farms had a positive operating profit margin. Small farms—in contrast—were less viable as businesses. Their average operating profit margin and rates of return on assets and equity were negative. Nevertheless, some farms in each small farm group had an operating margin of at least 20 percent.

A majority of each small farm group had a positive net farm income, but the average net income for each type of small farm was low compared with large- scale farms. Small farm households typically receive substantial off-farm income. Most off-farm income is from earned sources, from a wage and salary job or self-employment. Off-farm work dates back at least to the 1930s. The shift to full-time off-farm work, however, is more recent.

Because many farm households—particularly those operating small farms—are dual-career and receive a large share of their income from off-farm earnings, macroeconomic and monetary policies affecting the nonfarm economy are important to farm households. Also, a provision of the U.S. tax code allows farmers to write farm losses off against other income (Freshwater and Reimer, 1995, p. 220). This provision is especially important to operators of residential/lifestyle farms who have substantial off-farm earned income. Finally, the status of retirement programs is important to operators of retirement farms and to older operators in other farm types as they approach retirement.

Different Farms, Different Policies

Payments from commodity-related programs and conservation programs go to different types of farms. The distribution of commodity program payments is roughly proportional to the harvested acres of program commodities. As a result, medium-sales small farms and the two types of large-scale farms received 78 percent of commodity-related government payments in 2004. This report does not consider, however, how those payments are distributed for land-renting farm operators between land owners and operators.

In contrast, the Conservation Reserve Program (CRP), which pays the bulk of environmental payments, targets environmentally sensitive land rather than commodity production. As a result, retirement, residential/lifestyle, and low-sales small farms received 62 percent of conservation program payments in 2004. This distribution reflects the large number of farms in these groups, their large landholdings, and their tendency to enroll large shares of their land in the CRP. The program has relatively low labor and capital requirements, which make it attractive to residential/lifestyle farmers, who spend most of their work time off the farm, and to retired or older low- sales farmers, who have scaled back their operations.

A large majority of farms, 61 percent in 2004, do not receive government payments. Nevertheless, these farms—and the households that operate them—may be affected indirectly by government payments' impact on farmland and commodity markets. Some studies find that capitalizing government payments has increased farmland values by 15 to 25 percent in recent years (U.S. Dept. Agr., Office of the Chief Economist, 2003, p. 5). This would increase the net worth of landowing farm households, regardless of whether their farm received government payments. In addition, various analyses indicate that government payments have increased crop production between 1 and 6 percent over time (U.S. Dept. Agr., Office of the Chief Economist, 2003, p. 8). Thus, livestock producers who do not receive government payments may benefit from lower feed prices due to an increased supply of grain.

REFERENCES

Ahearn, Mary C., Janet E. Perry, and Hisham S. El-Osta. *The Economic Well-Being of Farm Operator Households, 1988-90.* AER-666. U.S. Dept. Agr., Econ. Res. Serv., January 1993.

Ahearn, Mary, Jet Yee, Eldon Ball, and Rich Nehring, with contributions by Agapi Somwaru and Rachel Evans. *Agricultural Productivity in the United States.* AIB-740. U.S. Dept. Agr., Econ. Res. Serv., January 1998.

Allen, Rich. "How to Interpret New Demographic Information in the Preliminary 2002 Census of Agriculture Release." Paper presented at the 2004 Agricultural Outlook Forum. Arlington, VA, February 19-20, 2004. (Available at: http://agecon.lib.umn.edu/cgi-bin/pdf_view.pl?paperid= 12779&ftype=.pdf.)

Bucks, Brian K., Arthur B. Kennickell, and Kevin B. Moore. "Recent Changes in U.S. Family Finances: Evidence from the 2001 and 2004 Survey of Consumer Finances," *Federal Reserve Bulletin,* March 22, 2006, pp. A1-A38. (Available at: http://www.federalreserve.gov/pubs/ bulletin/2006/financesurvey.pdf.)

Cash, A. James, II. "Where's the Beef? Small Farms Produce Majority of Cattle," *Agricultural Outlook*, U.S. Dept. Agr., Econ. Res. Serv., AGO- 297, December 2002, pp. 21-24.

Claassen, Roger, and Marc Ribaudo. "AREI Chapter 5.1: Conservation Policy Overview," in *Agricultural Resources and Environmental Indicators, 2006 Edition.* Keith Wiebe and Noel Gollehon, editors. EIB16. U.S. Dept. Agr., Econ. Res. Serv., July 2006.

Cochrane, Willard W. *The Development of American Agriculture: A Historical Analysis.* Minneapolis: University of Minnesota Press, 1993.

Freshwater, David, and Reimer, Bill. "Socio-economic Policies as Causal Forces for the Structure of Agriculture," in *Farms, Farm Families, and Farming Communities*, a special issue of the *Canadian Journal of Agricultural Economics*. David H. Harrington, Leslie Whitener, Ray D. Bollman, David Freshwater, and Philip Ehrensaft, editors. 1995, pp. 209-222.

Gale, Fred. "The Graying Farm Sector: Legacy of Off-Farm Migration," *Rural America,* U.S. Dept. Agr., Econ. Res. Serv., Vol. 17, Issue 3, Fall 2002, pp. 28-31.

Harrington, David H., Robert A. Hoppe, R. Neal Peterson, David Banker, H. Frederick Gale, Jr. "Changes in the Farm Sector," in *Financing Agriculture into the Twenty-first Century.* Marvin Duncan and Jerome M. Stam, editors. Boulder, CO: Westview Press, 1998.

Hellerstein, Daniel. "AREI Chapter 5.2: USDA Land Retirement Programs," in *Agricultural Resources and Environmental Indicators, 2006 Edition.* Keith Wiebe and Noel Gollehon, editors. EIB-16. U.S. Dept. Agr., Econ. Res. Serv., July 2006.

Hoppe, Robert A., editor. *Structural and Financial Characteristics of U.S. Farms: 2001 Family Farm Report.* AIB-768. U.S. Dept. Agr., Econ. Res. Serv., May 2001.

Hoppe, Robert A. "Farm Households Are Often Dual-Career," *Rural America*, U.S. Dept. Agr., Econ. Res. Serv., Vol. 16, Issue 2, Summer 2001, pp. 41-51.

Hoppe, Robert A., and David E. Banker. *Structure and Finances of U.S. Farms: 2005 Family Farm Report.* EIB-12. U.S. Dept. Agr., Econ. Res. Serv., May 2006.

Hoppe, Robert A., and Penni Korb. "Farm Numbers: Largest Growing Fastest," *Agricultural Outlook*, U.S. Dept. Agr., Econ. Res. Serv., AGO- 295, October 2002, pp. 24-27.

Jenkins, Warder B., and Hilton E. Robison. *Part-Time Farming in the United States.* Special Study, U.S. Census of Agriculture: 1935. U.S. Dept. Comm., Bur. of the Census, 1937.

Johnson, Cheryl D. *A Historical Look at Farm Income.* SB-807. U.S. Dept. Agr., Econ. Res. Serv., May 1990.

Lambert, Dayton, Patrick Sullivan, Roger Claassen, and Linda Foreman. *Conservation-Compatible Practices and Programs: Who Participates?* ERR- 14. U.S. Dept. Agr., Econ. Res. Serv., February 2006.

MacDonald, James, Janet Perry, Mary Ahearn, David Banker, William Chambers, Carolyn Dimitri, Nigel Key, Kenneth Nelson, and Leland Southard. *Contracts, Markets, and Prices: Organizing the Production and Use of Agricultural Commodities.* AER-837. U.S. Dept. Agr., Econ. Res. Serv., November 2004.

MacDonald, James M., and David E. Banker. "Agricultural Use of Production and Marketing Contracts," in *Structural and Financial Characteristics of U.S. Farms: 2004 Family Farm Report.* David E. Banker and James M. MacDonald, editors. AIB-797. U.S. Dept. Agr., Econ. Res. Serv., March 2005.

McBride, William D., and Nigel Key. *Economic and Structural Relationships in U.S. Hog Production.* AER-818. U.S. Dept. of Agr., Econ. Res. Serv. February 2003.

Mishra, Ashok K., Ron L. Durst, and Hisham S. El-Osta. "How Do U.S. Farmers Plan for Retirement?" *Amber Waves*, U.S. Dept. Agr., Econ. Res. Serv., Vol. 3, Issue 2, April 2005, pp. 12-18.

Perry, Janet E., and Mary C. Ahearn. *Limited Opportunity Farm Households in 1988.* AIB-662. U.S. Dept. Agr., Econ. Res. Serv., February 1993.

Reimund, Donn A., and Fred Gale. *Structural Change in the U.S. Farm Sector, 1974 -87: 13th Annual Family Farm Report to Congress.* AIB647. U.S. Dept. Agr., Econ. Res. Serv., May 1992.

U.S. Department of Agriculture, National Agricultural Statistics Service. *1997 Census of Agriculture*, Vol. 1: Geographic Area Series, Part 51: United States Summary and State Data, AC97-A-5 1. March 1999.

U.S. Department of Agriculture, National Agricultural Statistics Service. *2002 Census of Agriculture*, Vol. 1: Geographic Area Series, Part 51: United States Summary and State Data, AC-02-A-51. June 2004.

U.S. Department of Agriculture, National Agricultural Statistics Service. *2004 Agricultural Resource Management Survey Phase III Survey Administration Manual*. January 2005.

U.S. Department of Agriculture, National Commission on Small Farms. *A Time to Act: A Report of the USDA National Commission on Small Farms.* Miscellaneous Publication 1545 (MP-1545). January 1998.

U.S. Department of Agriculture, Office of the Chief Economist. *Report of the Commission on the Application of Payment Limitations for Agriculture.* August 2003.

U.S. Department of Commerce, Bureau of the Census. *Historical Statistics of the United States, Colonial Times to 1970, Bicentennial Edition.* 1975.

U.S. Department of Labor, Bureau of Labor Statistics. "Revisions to the Current Population Survey effective in January 2003," *Employment and Earnings*, Vol. 50, No. 2, February 2003, pp. 4-23.

U.S. Department of Labor, Bureau of Labor Statistics. Household data, annual averages. *Employment and Earnings*, Vol. 52, No. 1, January 2005.

U.S. National Archives and Records Administration, Office of the Federal Register. *Federal Register.* Vol. 68, No. 104, May 30, 2003.

U.S. Office of the President. *Economic Report of the President.* Transmitted to the Congress, together with the Annual Report of the Council of Economic Advisors, February 2006.

APPENDIX 1

Measuring Operator Household Income and Net Worth

The Current Population Survey (CPS), conducted by the Bureau of the Census, is the source of official U.S. household income statistics. Thus, calculating an estimate of farm household income from the Agricultural Resource Management Study (ARMS) that is consistent with CPS methodology allows income comparisons between farm operator households and all U.S. households.

The CPS definition of farm self-employment income is net money income from the operation of a farm by a person on his or her own account, as an owner or renter. CPS self-employment income includes income received as cash, but excludes in-kind or nonmoney receipts. No adjustments are made to the CPS income measure to reflect inventory changes, since inventory change is a nonmoney item. The CPS definition departs from a strict cash concept by deducting depreciation, a noncash business expense, from the income of self-employed people.

Farm self-employment income from the ARMS is the sum of the share of farm business income (net cash farm income less depreciation) accruing to the principal operator's household, wages paid to the operator, and net rental income from renting farmland. Adding other farm-related income of the operator household yields earnings of the operator household from farming activities. (Other farm-related earnings consist of net income from a farm business other than the one being surveyed, wages paid by the farm business to household members other than the operator, and commodities paid to household members for farm work.)

ARMS is also the source of data for estimates of operator households' net worth. The net worth of farm operator households is defined as the difference between their assets and liabilities. It is calculated as the sum of the operator household's farm net worth and nonfarm net worth. If the net worth of the farm is shared with other households (such as the households of shareholders in a family corporation), only the operator household's share is included.

For more information on operator household income and well-being, see "Farm Household Economics and Well-Being," a briefing room on the ERS website (http://www.ers.usda.gov/Briefing/WellBeing/). Estimates presented in this report are consistent with those from the briefing room. Both sets of estimates are derived from ARMS for the principal operator households using CPS procedures. Household income estimates cannot be derived from the sector estimates of net farm income presented on another ERS briefing room, "Farm Income and Costs" (http://www.ers.usda.gov/Briefing/FarmIncome/). The farm sector estimates are estimated from several data sources and include all participants in farm production, including contractors and share landlords who do not farm. For more information, see Harrington et al. (1998, pp. 45-52).

APPENDIX II

Defining Limited-Resource Farms—Past, Present, and Future

Perry and Ahearn (1993) first identified limited-resource farmers for the Economic Research Service (ERS)—based on 1988 data—although they used the term "limited-opportunity" rather than "limited-resource." They defined limited-resource farms using three criteria:

Farm sales less than $100,000
Farm assets less than $150,000
Operator household income less than the poverty level

When the Economic Research Service created its farm classification system in 1998, it incorporated the Perry-Ahearn definition, with one modification. Family income was required to be below $20,000 rather than the poverty level. Using a $20,000 cutoff rather than the poverty level avoided the necessity of knowing family size. Family size is not collected every year by the Agricultural Resource Management Survey (ARMS), but it is needed to assign the appropriate poverty level to a family (Hoppe, 2001, p. 4).

The Current Definition

In 2003, a new definition of limited-resource farms was developed by an interagency committee to provide a consistent definition across all USDA agencies (U.S. National Archives and Records Admn., 2003, p. 32520). This USDA-wide definition is currently used in the ERS farm classification. The limits on sales and household income are similar under the former and current definitions (see box, "Defining Limited-Resource Farms"). Both definitions use a $100,000 cutoff for farm sales, although the current definition indexes the cutoff to reflect price changes. The current definition also requires 2 years of low sales, rather than 1. The cutoff for household income is also set low in both definitions, but—as in the case of sales—2 years of low income are required under the current definition.

The main difference between the two definitions is the absence of a limit on farm assets in the current definition. An asset limitation was not used because the assets held by individual farmers are difficult to verify on applications to participate in USDA programs targeted at limited-resource farmers. Instead, the requirement for a second year of low income—which is easier to verify than low assets—was added as an indication of persistently low income.

Differences

The limited-resource farms identified under the two definitions are different in some respects (app. table 1). Although the level of household income is similar under the two definitions, median farm assets are nearly three times as high under the current definition ($244,600) as under the former definition ($87,600). Similarly, the net worth of farm households is nearly three times higher under the current definition. Operators under the current definition also tend to be older: 65, on average, under the current definition versus 61 under the former definition.

A total of 197,700 farms were classified in 2004 as limited-resource farms under the current definition. This includes 56,300 farms that were also classified as limited-resource under the former definition (appendix figure 1). The 141,400 farms added under the new definition have a higher level of farm assets—a median of $334,200—compared with farms classified as limited-resource under the former definition. Operators of the added farms also tend to be older.

Defining Limited-Resource Farms

Criterion	Former definition in the ERS farm classification	Current definition (USDA-wide)
Sales	Less than $100,000, with no indexing	Low sales in both the current and previous year. Low sales is defined as less than $100,000 in 2003 and indexed thereafter
Farm assets	Less than $150,000	No limits
Operator household income	Less than $20,000 in the current year	Low in both the current and previous year. Income is low if it is less than the poverty level for a family of four with two children— $19,157 in 2004— or if it is less than half the county median household income

Appendix table 1. Characteristics of limited-resource farmers under the former and current definitions, 2004

Item	Former definition	Current definition
	Number	
Farms and operator households	74,819	197,734
Dollars per household (or farm)		
Median household income (2004)	9,900	10,300
Median farm assets	87,614	244,609
Median household net worth	109,463	271,280
Years		
Average age	61	65

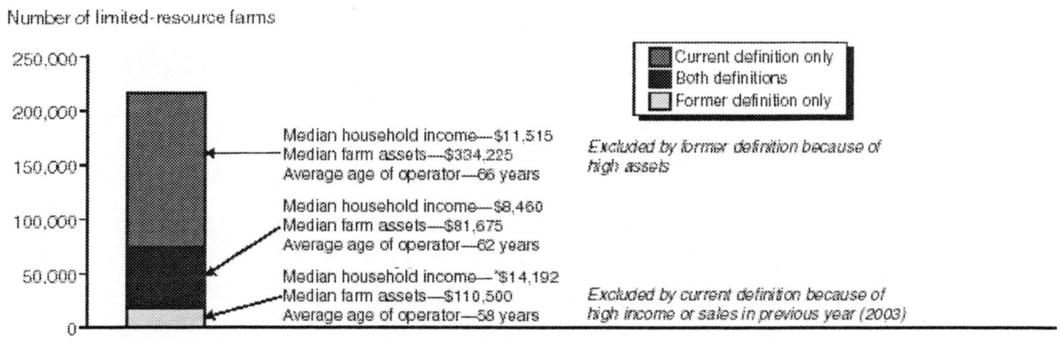

Source: USDA, Economic Research Service, 2004 Agricultural Resource Management Survey, Phase III.

Appendix figure 1. **Limited-resource farms under the former and current definitions, 2004**
Limited-resource farms added by the current definition have more assets and older operators

Alternate Definitions

The current limited-resource definition focuses on low-sales farms operated by farmers with low household income over a 2-year period. Different criteria could be considered, at least for research purposes. For example, we could use a definition that included an asset constraint, such as the previous $150,000 limit. We may want to even consider an asset constraint that changes over time—such as one-half of the median assets of all small farms—to reflect increases in the value of farmland and other assets used in farming. Different sales constraints might also be tested.

In future *Family Farm Reports*, ERS will continue to provide information about limited-resource farms as defined currently because that definition is used by USDA agencies to administer programs. We will also explore alternate definitions to identify other groups of limited-resource farms. To facilitate comparing the USDA-wide current definition with alternate definitions, the limited-resource category will be dropped from the ERS farm classification system in future reports. However, there will be more information on counts and characteristics of limited-resource farmers under different definitions.

End Notes

[1]Differences between ARMS-based estimates are stressed in this report only when the estimates are significantly different at the 95-percent confidence level or more.

[2]Nonfamily farms' share of production increased by 1.5 percentage points between 2003 and 2004 (from 13.7 percent to 15.2 percent). This change, however, was not statistically significant.

[3]ARMS collects the number of hours worked on farm by the principal operator, the spouse of the principal operator, other operators, and unpaid workers. The survey does not collect hours worked by hired or contract labor, however. Hours of hired and contract labor are estimated by dividing hired labor and contract labor expense by the State-specific wage rate for farm labor.

[4]Traditionally, farm data sources in the United States assumed each farm had only one operator. The "one farm, one operator" assumption was dropped when the census of agriculture and ARMS collected data for 2002. Both the census and ARMS now count all operators—principal and secondary— and ask for detailed information on up to three operators.

[5]Approximately 17 percent of the operators of nonfamily farms were women, but the difference between that estimate and the 9-percent estimate for all farms is not statistically significant.

[6]In the late 1980s, ERS developed a measure of financial position that considered both income and solvency. Under this classification system, farms were classified as being in one of four financial categories: favorable, marginal-income, marginal-solvency, or vulnerable. For definitions of the four categories, see footnote 6 in table 5.

[7] ERS plans to explore alternate limited-resource definitions to identify different types of limited-resource farms and farmers. It will continue to provide information about limited-resource farms as defined under the current definition, because that definition is used by USDA agencies to administer programs. To facilitate comparing the current and alternate definitions, the limited-resource category will be dropped from the ERS farm classification system. However, there will be more information on counts and characteristics of limited-resource farmers under different definitions.

[8]ARMS data rely entirely on the respondent for program-related information. As a result, the survey shows different levels and composition of government payments than do administrative data, which are based on payment records kept by the agencies involved. In addition, ARMS records only the payments received by farmers, while the administrative data include payments received by persons who do not farm, mainly nonoperator landlords.

[9] If a place does not have $1,000 in sales, a "point system" assigns values for acres of various crops and head of livestock to estimate normal sales. "Point farms" are farms with less than $1,000 in sales but points worth at least $1,000. See "What is the Definition of a Farm?" on the NASS website (http://www.nass.usda.gov/Census_of_Agriculture/Frequently_Asked_Questions/index. asp#1).

[10]Enrollment in the CRP began in 1986 and enrollment in the WRP began in 1992 (Hellerstein, 2006). Since neither program existed in 1982, the farm count from the 1982 census and the farm counts from the 1997 and 2002 censuses are comparable, as far as the treatment of CRP/WRP farms is concerned.

[11]Undercoverage is much less an issue for sales than for the farm count. The five censuses prior to 2002 included an average of 92 percent of farms but 98 percent of production (USDA, NASS, 1999, p. C-5).

[12]This measure is the median of acres harvested by enterprise size, not the median of farms by enterprise size. Under the latter method, farms would be arrayed by acres harvested and the median divides *farms* into two equal groups, not the acres harvested. By using acres harvested, our definition of median identifies the enterprise size at the midpoint of enterprises arrayed by a measure of production.

In: America's Family Farms
Editor: Efren J. Tamayo
ISBN: 978-1-60741-751-4

Chapter 10

U.S. FARM INCOME*

Randy Schnepf

SUMMARY

Despite high production costs, 2008 represented another year of record profitability for the U.S. farm economy as a whole. According to USDA's Economic Research Service (ERS), national net farm income—a key indicator of U.S. farm well-being—rose to a record $86.9 billion in 2008, marginally above the previous year's record ($86.8 billion). The growth in cash receipts to a record $323.4 billion for crop and livestock sales (up $38.6 billion or 14% from 2007) was nearly offset by a surge in production costs (up $38.2 billion or 15%) to a record $292.5 billion.

Government farm payments are projected up slightly in 2008 at $12.5 billion. An increase in ad hoc disaster payments more than offset lower commodity program payments, which declined when high crop prices rose above the price triggers for marketing loan benefits and countercyclical payments in 2008.

Within the farm balance sheet, total farm asset value of $2,359 billion and total farm debt of $212 billion are both projected at record levels in 2008. The debt-to-asset ratio of 9.0% is down sharply from last year's value of 9.6% and represents the lowest level since 1960, suggesting a strong financial position for the agricultural sector as a whole.

However, less than ideal market conditions heading into 2009 suggest dim prospects for the longer-term farm income outlook, albeit surrounded by considerable uncertainty. On the one hand, the global financial crisis, economic recession, rising unemployment, limited credit availability, and plummeting asset values that persist in early 2009 have contributed to substantial "demand destruction" (i.e., a severe weakening of consumer demand), which bodes poorly for farm commodity price prospects. On the other hand, weak energy markets and declining input prices could provide some spark to both producer investment and

---

* This is an edited, reformatted and augmented version of a CRS Report for Congress publication dated January 2009.

consumer demand for agricultural sector products, perhaps by the middle to latter half of the year. USDA will release its first U.S. farm income forecasts for 2009 on February 12, 2009.

This report supersedes CRS Report RS21970, *The U.S. Farm Economy*, by Randy Schnepf. It will be updated as events warrant.

INTRODUCTION

The U.S. farm sector is vast and varied. It encompasses production activities related to traditional field crops (such as corn, soybeans, wheat, and cotton), livestock and poultry products (including meat, dairy, and eggs), as well as fruits, tree nuts, and vegetables. In addition, U.S. agricultural output includes greenhouse and nursery products, forest products, custom work, machine hire, and other farm-related activities. The intensity and economic importance of each of these activities, as well as their underlying market structure and production processes, vary regionally based on the agro-climatic setting, market conditions, and other factors. As a result, farm income and rural economic conditions may vary substantially across the United States.[1] However, this report focuses singularly on aggregate national net farm income and the farm debt-to-asset status as reported by the U.S. Department of Agriculture (USDA).[2]

Annual U.S. net farm income is the single most watched indicator of farm sector well-being, as it captures and reflects the entirety of economic activity across the range of production processes, input expenses, and marketing conditions that have persisted during a specific time period. When national net farm income is reported together with a measure of the national farm debt-to-asset situation, the two summary statistics provide a quick indicator of the economic well-being of the national farm economy.

Two different indicators measure farm profitability: net cash income and net farm income.

Net cash income compares cash receipts to cash expenses. As such, it is a cash flow measure representing the funds that are available to farm operators to meet family living expenses and make debt payments. For example, crops that are produced and harvested but kept in on-farm storage are not counted in net cash income.

Net farm income is a value of production measure, indicating the farm operator's share of the net value added to the national economy within a calendar year, independent of whether it is received in cash or noncash form. In contrast to net cash income, net farm income includes the value of home consumption, changes in inventories, capital replacement, and implicit rent and expenses related to the farm operator's dwelling that are not reflected in cash transactions during the current year.

Net cash income is generally less variable than net farm income. Farmers can manage the timing of crop and livestock sales and of the purchase of inputs to stabilize the variability in their net cash income. For example, farmers can hold crops from large harvests to sell in the forthcoming year, when output may be lower and prices higher. Off-farm income, which has increased in importance in recent decades, is not included in the calculation of aggregate farm income. Instead, it is included in the discussion of farm income at the household level.

REVIEW OF CELENDER YEAR 2008

Despite the dramatic price declines that occurred during the second half of 2008 (**Figure 2** and **Figure 3**), USDA's net farm income estimate for 2008 remains at the historically high level of $86.9 billion, up marginally from the previous year's record $86.8 billion (**Table 1**).[3] When measured in cash terms, net cash income in 2008 is projected up nearly 4% to $90.7 billion, compared with the previous year's record of $87.4 billion (**Figure 1**). Net cash income is projected to rise more than net farm income because of the carryover of unsold crops from 2007 for sale in 2008.

From a historical perspective, the six years extending from 2003 through 2008 represent the six highest years for U.S. farm income on record. National net cash income averaged nearly $81 billion per year during that six-year period, well above the previous single year high of $60.8 billion achieved first in 1993 and again in 1997. Prospects for 2009 are less sanguine, as commodity prices continue to weaken for most major field crops and livestock products heading into 2009 (**Figure 2** and **Figure 3**). Only rice and broilers appear to have reversed, at least temporarily, the downward trend.

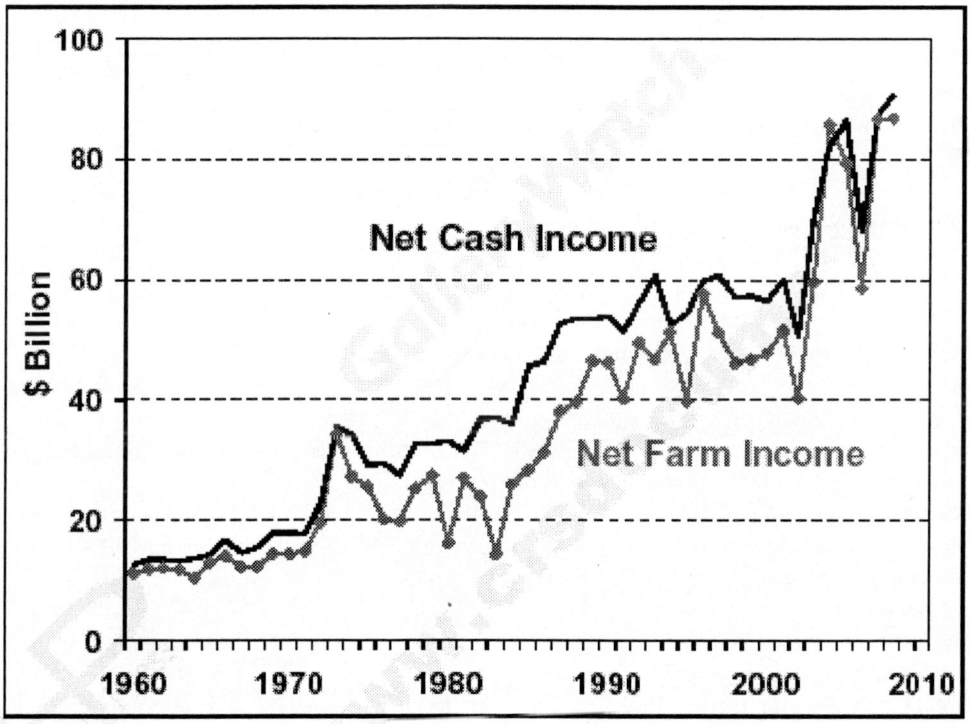

Source: USDA, Economic Research Service, "2008 Farm Income Forecast," November 25, 2008, at http://www.ers.usda.gov/Briefing/FarmIncome/.

Notes: All values are in nominal terms, i.e., they are not adjusted for inflation. 2008 is projected.

Figure 1. Annual U.S. Farm Sector Income, 1960 to 2008F

Table 1. Annual U.S. Farm Income Since 2002 (($ billions))

Item	2002	2003	2004	2005	2006	2007	2008[a]
1. Cash receipts	195.0	215.6	237.2	240.9	240.8	284.8	323.4
Crops[b]	101.0	109.9	113.6	116.0	122.6	147.0	179.9
Livestock	94.0	105.6	123.6	124.9	118.2	137.9	143.5
2. Government payments[c]	12.4	16.5	13.0	24.4	15.8	11.9	12..5
Fixed direct payments[d]	3.9	6.4	5.2	5.2	5.1	5.1	5.2
CCP[e]	0.2	2.3	1.1	4.1	4.0	1.1	0.7
Marketing Loan Benefits[f]	2.8	1.3	3.5	7.1	1.8	1.1	0.1
Conservation	2.0	2.2	2.3	2.8	3.0	3.1	3.2
Ad hoc and emergency	1.7	3.1	0.6	3.2	0.3	0.5	2.7
All other[g]	1.9	1.2	0.2	2.1	1.7	1.0	0.6
3. Farm-related income[h]	14.8	15.7	17.1	16.2	17.5	16.6	17.6
4. Gross cash income (1+2+3)	222.2	247.8	267.3	281.5	274.1	313.4	353.5
5. Cash expenses[i]	171.6`	177.6	185.0	194.8	206.0	226.0	262.8
6. NET CASH INCOME	50.7	70.2	82.3	86.6	68.0	87.4	90.7
7. Total gross revenues[j]	233.6	260.0	295.6	301.1	292.4	341.1	379.4
8. Total production expenses[k]	193.4	200.3	209.8	221.8	233.9	254.4	292.5
9. NET FARM INCOME	40.2	59.7	85.8	79.3	58.5	86.8	86.9

Source: USDA, Economic Research Service, briefing rooms: *Farm Income and Costs: Farm Sector Income*, and Costs: Farm Sector Income, available at http://www.ers.usda.gov/ Briefing/FarmIncome/; U.S. farm income data updated as of November 25, 2008.

a. Data for 2008 are USDA forecasts.

b. Includes Commodity Credit Corporation loans under the farm commodity support program.

c. Government payments reflect payments made directly to all recipients in the farm sector, including landlords. The non-operator landlords' share is offset by its inclusion in rental expenses paid to these landlords and thus is not reflected in net farm income or net cash income. For more information on U.S. farm commodity programs, see CRS Report RL34594, Farm Commodity Programs in the 2008 Farm Bill, by Jim Monke; for more information on conservation programs see CRS Report RL34557, Conservation Provisions of the 2008 Farm Bill, by Tadlock Cowan, Renée Johnson, and Megan Stubbs.

d. Direct payments include production flexibility payments of the 1996 Farm Act through 2001, and fixed direct payments under the 2002 Farm Act since 2002.

e. CCP = counter-cyclical payments.

f. Includes loan deficiency payments (LDP); marketing loan gains (MLG); and commodity certificate exchange gains.

g. Peanut quota buyout, milk income loss payments, and other miscellaneous program payments.

h. Income from custom work, machine hire, agri-tourism, forest product sales, and other farm sources.

i. Excludes depreciation and perquisites to hired labor.

j. Gross cash income plus inventory adjustments, the value of home consumption, and the imputed rental value of operator dwellings .

k. Cash expenses plus depreciation and perquisites to hired labor.

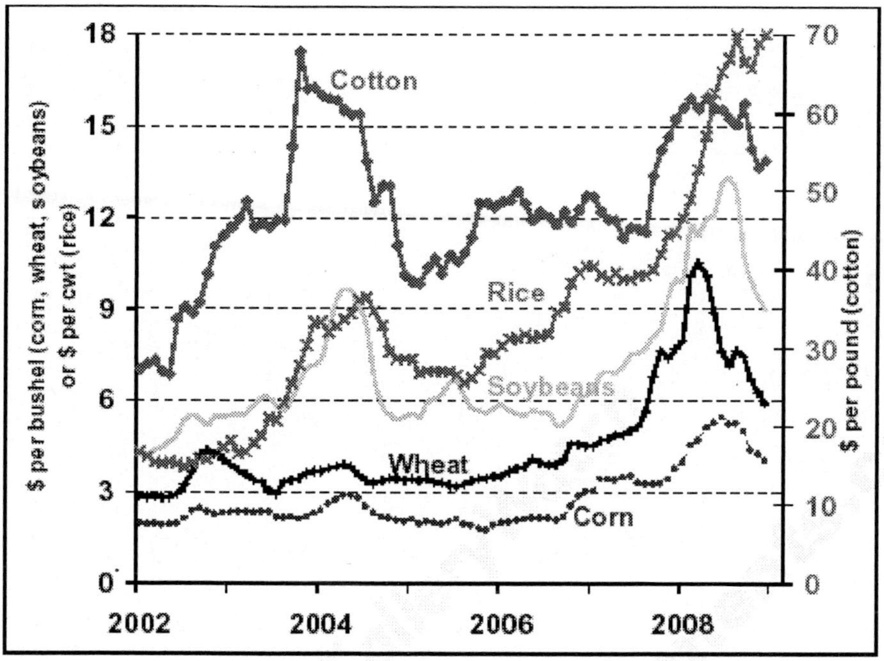

Source: USDA, National Agricultural Statistics Service.

Figure 2. Monthly Farm-Prices-Received for Major Field Crops

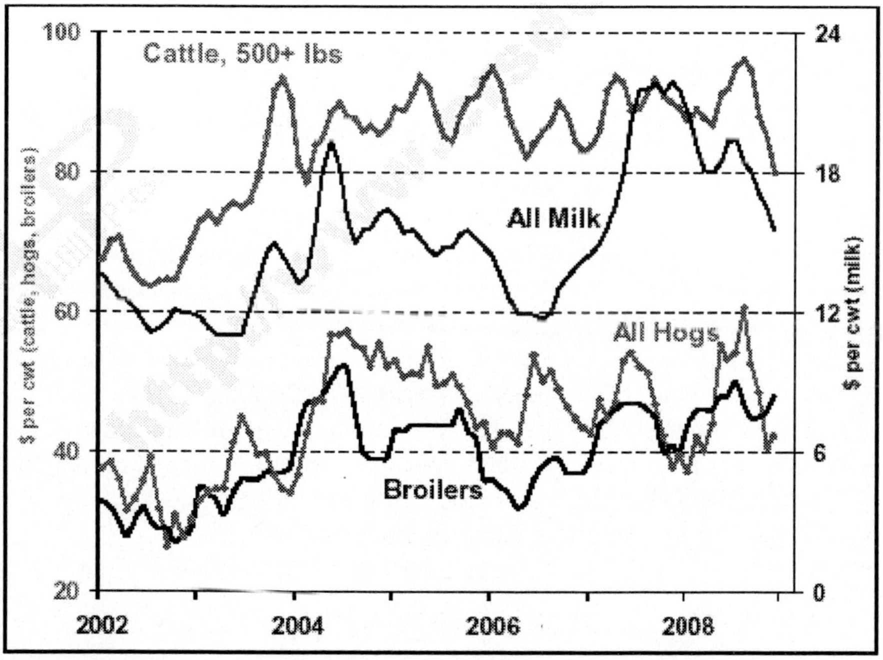

Source: USDA, National Agricultural Statistics Service.
Note: cwt = hundredweight or units of 100 lbs.

Figure 3. Monthly Farm-Prices-Received for Major Livestock Products

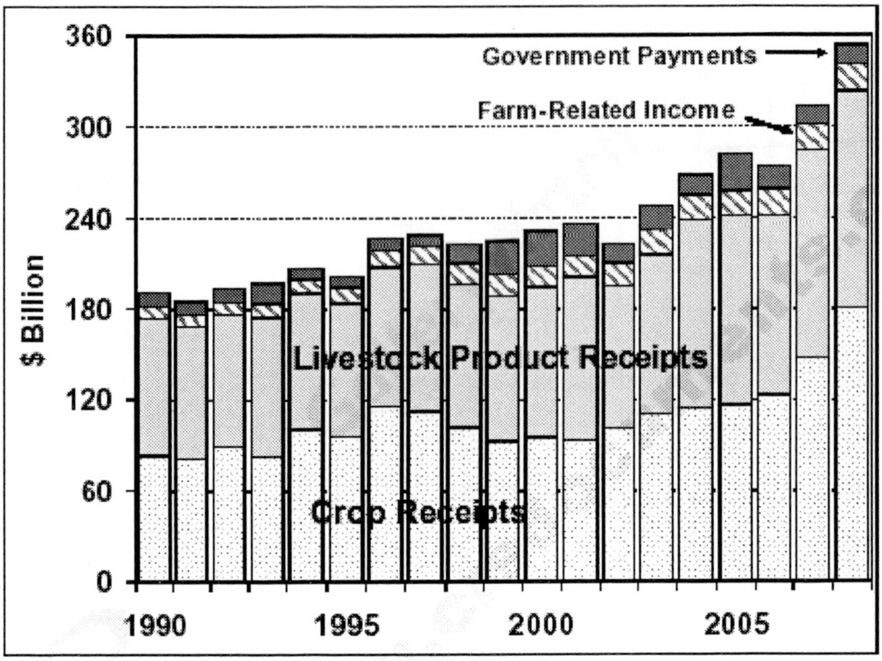

Source: USDA, Economic Research Service, "2008 Farm Income Forecast," November 25, 2008, at http://www.ers.usda.gov/Briefing/FarmIncome/.

Notes: 2008 is projected. Receipts from crop and livestock product sales, as well as government payments, are described in more detail below. Farm-related income includes income from custom work, machine hire, agritourism, forest product sales, insurance indemnities, and cooperative patronage dividend fees.

Figure 4. Farm Cash Receipts by Source

CASH RECEIPTS

Good harvests, strong prices, and robust domestic and international demand combined to push farm gross receipts to record levels in 2008. The combined value of cash receipts from sales of both crop and livestock commodities is projected at $323.4 billion in 2008, the highest amount on record—up $38.6 billion from the previous year's record of $284.9 billion—and driven almost entirely by higher crop prices (**Table 1** and **Figure 4**).

In recent years, U.S. domestic demand has been supported by the rapid emergence of the U.S. agriculture-based ethanol industry. In addition, strong export demand through 2007 and the first half of 2008, aided in part by a weak dollar, helped to draw stocks for major grains and oilseeds to historically low levels in 2008, thus supporting higher market prices. While crop farmers rejoiced, livestock feeders expressed concern about the escalating costs of feed—the largest single cost component for cattle, dairy, hog, and poultry production.

CROPS

Strong commodity prices, although highly volatile since mid-2008, are expected to generate record crop cash receipts in 2008 of $180 billion (**Figure 5**). Sales of field crops (i.e., feed, food, and oil crops) increased over 29% from 2007 to $113.3 billion. This total includes feed crop (i.e., corn, sorghum, barley, and oats) sales of $60.2 billion, up 42%, food crop (i.e., wheat and rice) sales of $19.2 billion, up over 50%, and oil crop (i.e., soybeans, sunflowers, rapeseed/canola, and other minor oilseeds) sales of $28.7 billion, up 27%. Corn cash receipts alone are projected at a record $50 billion in 2008—the single most valuable farm crop ever harvested in the United States. Cash receipts for soybeans ($26.6 billion), wheat ($16.1 billion), and rice ($3.1 billion) also are expected to establish single-season sales value records. Cotton was the most notable deviation from this trend, with a sales decline of 17% to $5.1 billion, due to sharply lower plantings and output in 2008.

Crop-specific output and sales are increasingly being influenced by the rapid expansion of cornbased biofuels production, due in large part to strong federal incentives that include a blending tax credit, a usage mandate, and a prohibitive import tariff on foreign-produced biofuels. With this strong support, the U.S. corn-based ethanol industry has grown rapidly since mid-2004, when production capacity was estimated at around 3 billion gallons per year, to over 10.8 billion gallons as of December 30, 2008.[4] The U.S. ethanol sector received a substantial boost in December 2007 when the Energy Independence and Security Act (EISA) was signed into law (P.L. 110-140). EISA greatly expands the mandate for corn-based ethanol use from 4.7 billion gallons in 2007 to 9 billion in 2008 and 15 billion by 2015.[5] USDA estimates that nearly 30% of the 2008 corn crop will be used to produce ethanol during the 2008/2009 (September-August) corn marketing year.[6] This additional demand has helped to push corn and other crop prices steadily higher since 2005 as they compete for a fixed amount of cropland (**Table 4**).

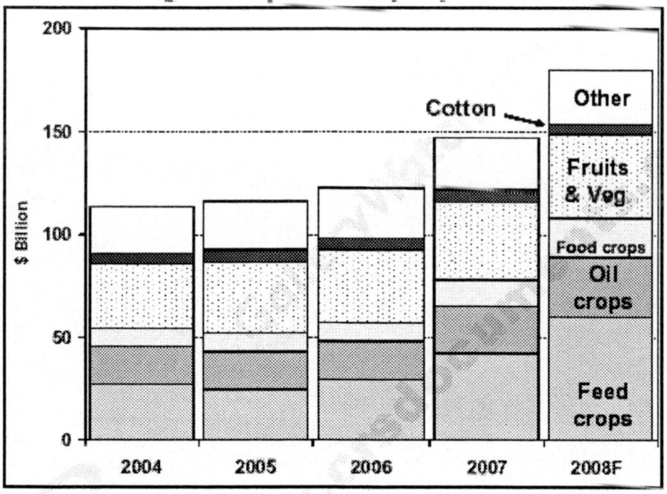

Source: USDA, Economic Research Service, "2008 Farm Income Forecast," November 25, 2008, at http://www.ers.usda.gov/Briefing/FarmIncome/.

Notes: 2008 is projected.

Figure 5. Crop Cash Receipts by Source

Livestock

As with crop sales, the value of livestock product sales also is forecast record high in 2008 at $143.5 billion, up 4% from the previous year's record of $137.9 billion (**Figure 6**).

Strong market prices for most major livestock categories—beef, poultry, and hogs—during 2008 were the driving factor behind record cash receipts for livestock products. However, price prospects have weakened considerably heading into 2009—milk prices are projected about 34% lower in 2009 and eggs down about 4% (**Table 4**).

Government Payments

Direct government payments are forecast at $12.5 billion in 2008, up about 5% from $11.9 billion in 2007 but well below the record of $24.4 billion in 2005. Higher market prices during 2008 limited payments under the two major price-contingent programs—counter-cyclical payments (CCP), which are projected to decline to about $0.7 billion, and marketing loan benefits (loan deficiency payments, marketing loan gains, and certificate exchange gains), which are projected to fall to about $90 million in 2008.[7] In contrast, farm disaster assistance—which (along with other emergency assistance) has figured heavily in farm sector income in most of the previous 20 years (1989-2008)[8]—is projected up at $3.2 billion in 2008.

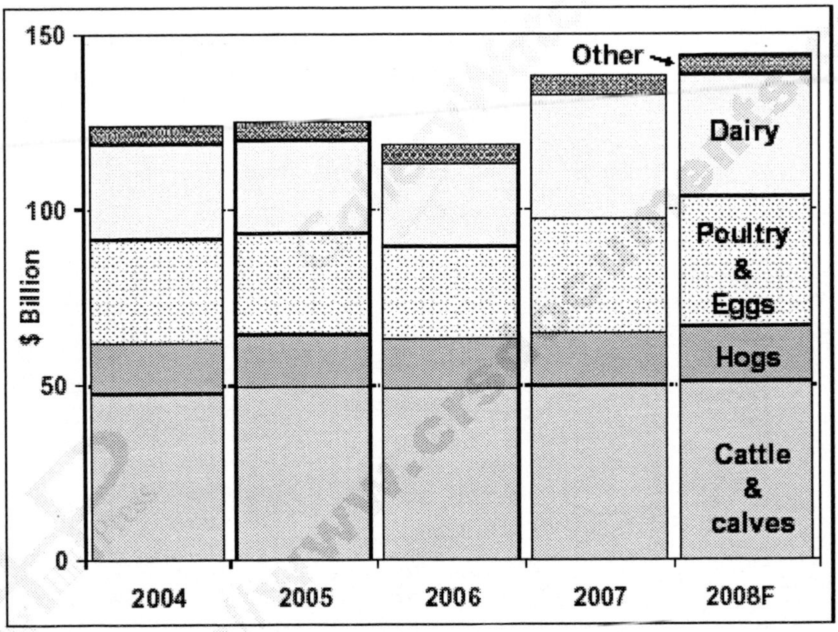

Source: USDA, Economic Research Service, "2008 Farm Income Forecast," November 25, 2008, at http://www.ers.usda.gov/Briefing/FarmIncome/.

Notes: 2008 is projected.

Figure 6. U.S. Livestock Product Cash Receipts by Source

Fixed direct payments, whose payment rates are fixed in legislation and are not affected by the level of program crop prices, have been $5.1 to $5.2 billion since 2004 and will only vary based on payment timing and participation rates. Conservation payments have grown slowly but steadily since 1998, and are expected to reach $3.2 billion in 2008.

Production Expences

An emerging concern for the U.S. farm sector has been the rapid escalation of total farm production expenses (**Figure 8** and **Figure 9**), which have risen an average of 7.8% annually since 2003 compared with an average annual growth rate of only 7.1% for gross cash income.

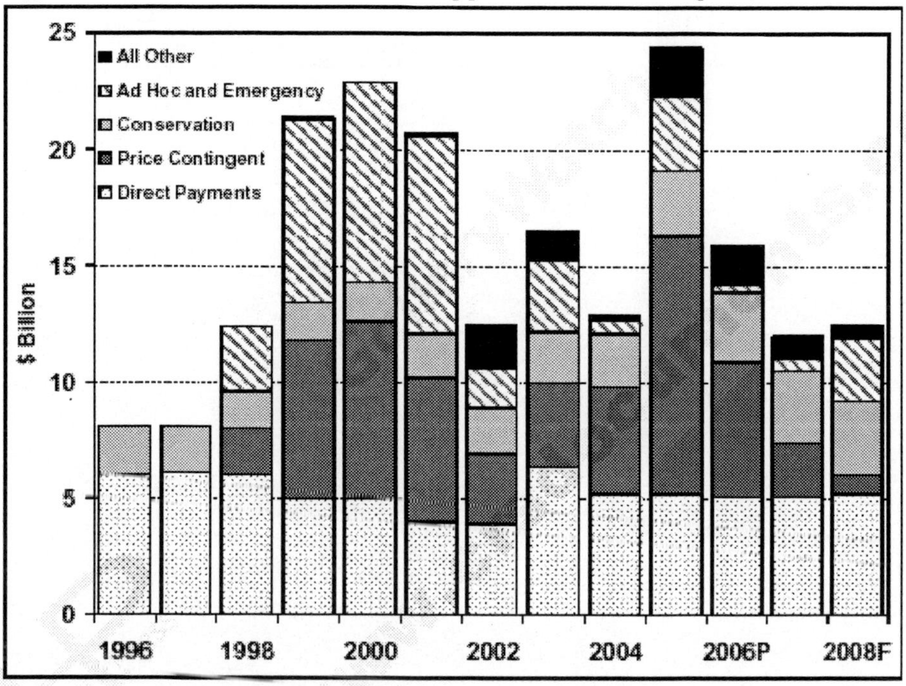

Source: USDA, Economic Research Service, "2008 Farm Income Forecast," November 25, 2008, at http://www.ers.usda.gov/Briefing/FarmIncome/; and USDA, Risk Management Agency, Current Year-to-Date National Summary of Business Reports, at http://www.rma.usda.gov/data/sob.html.

Notes: Data are on a fiscal year basis and may not correspond exactly with the crop or calendar year; 2008 is projected. Direct payments include production flexibility contract payments enacted under the 1996 farm bill and fixed direct payments of the 2002 and 2008 farm bills; price-contingent outlays include loan deficiency payments (LDPs), marketing loan gains, and counter-cyclical payments (CCPs); conservation outlays include Conservation Reserve Program (CRP) payments along with other conservation program outlays; Ad Hoc and Emergency includes emergency supplemental crop and livestock disaster payments and market loss assistance payments for relief of low commodity prices; and "all other" outlays include peanut quota buyout payments, milk income loss payments, tobacco transition payments, and other miscellaneous expenditures.

Figure 7. U.S. Government Farm Support, Direct Outlays, 1996 to 2008F

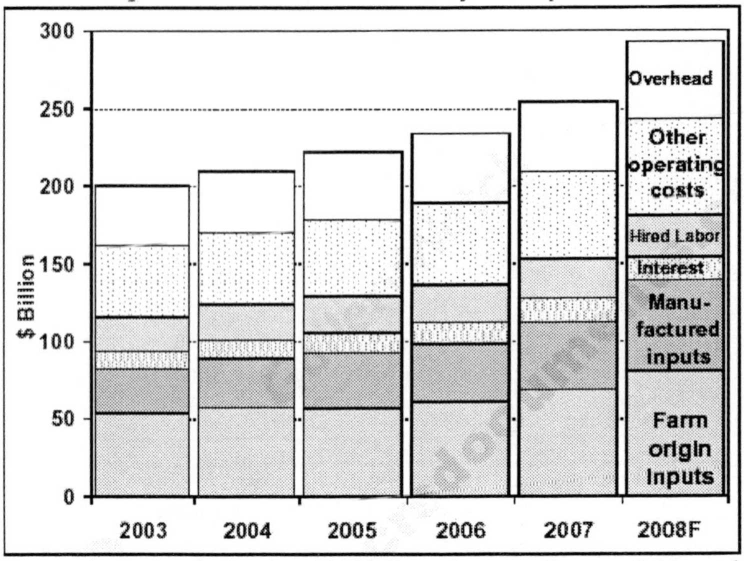

Source: USDA, Economic Research Service, "2008 Farm Income Forecast," November 25, 2008, at http://www.ers.usda.gov/Briefing/FarmIncome/.

Notes: 2008 is projected. Farm origin inputs include purchases of feed, livestock and poultry, and seed. Manufactured inputs include fertilizers and lime, pesticides, petroleum fuel and oils, and electricity. Other operating costs include repair and maintenance of capital items, machine hire and custom work, marketing storage, transportation expenses, and other miscellaneous expenses. Overhead expenses include property taxes, net rent to a non-operator landlord, and capital consumption.

Figure 8. Farm Cash Production Expenses by Source

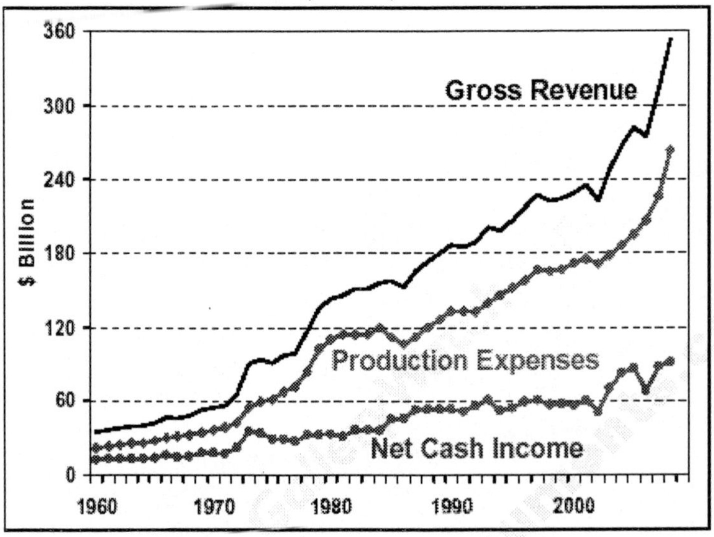

Source: USDA, Economic Research Service, "2008 Farm Income Forecast," November 25, 2008, at http://www.ers.usda.gov/Briefing/FarmIncome/.

Notes: All values are in nominal terms, i.e. not adjusted for inflation. 2008 is projected.

Figure 9. U.S. Farm Gross Revenue, Production Expenses, and Net Income (cash basis)

Total production expenses are forecast at a nominal record $262.8 billion in 2008, up $36.8 billion (16%) from 2007's record level (**Figure 8**). Higher prices for feed crops (although beneficial for feed crop producers) have raised feed costs nearly 23% year-to-year for livestock producers. Feed costs are expected to rise to $47 billion in 2008, while escalating energy costs push fuels and fertilizers up 26% and 64% year-over-year, respectively, to record outlays of $16.4 and $27.5 billion in 2008. However, both feed and energy costs declined sharply in the last quarter of 2008, and appear likely to remain at substantially lower levels in 2009.

AVERAGE FARM HOUSEHOLD INCOME

Average farm household income is projected at a record $86,798 in 2008, up 0.7% from the previous year's record. The share of farm income derived from off-farm sources has been increasing steadily in recent decades. In 2008, off-farm income sources account for over 93% of the national average farm household income, compared with less than 7% from farming activities (**Figure 10**).

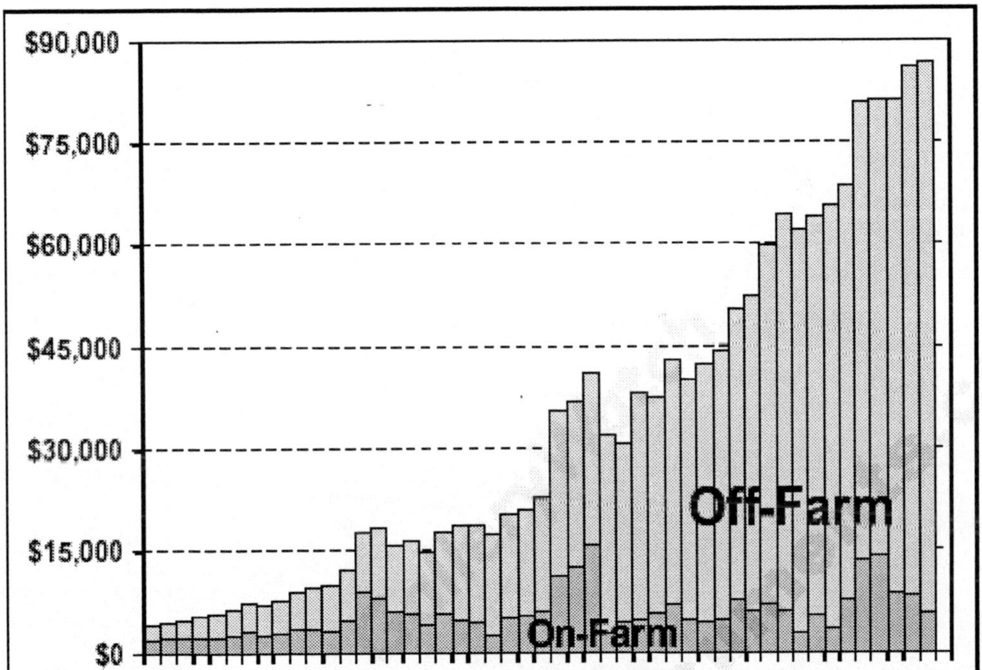

Source: USDA, Economic Research Service, "2008 Farm Income Forecast," November 25, 2008, at [http://www.ers.usda.gov/Briefing/FarmIncome/].

Notes: 2008 is projected.

Figure 10. U.S. Average Farm Household Income, by On- and Off-Farm Sources

The share of income from farming increases with farm size (as measured by gross sales). "Large" commercial farm households (farms with annual sales between $250,000 and $499,999), on average, obtained 60% of their total household income from farming activities in 2007, while "very large" family farms (farms with annual sales in excess of $500,000) obtained nearly 80% of household income on-farm.9 These two classes of farms represented slightly less than 8% of family farms. Intermediate family farms (farms with annual sales in excess of $100,000 but less than $250,000) represented about 28% of family farms and obtained about 32% of household income from on-farm sources. The remaining 64% of family farms are classified as rural residence farms and either receive little or no income from farm sources or have a total income level that qualifies them as limited-resource farms.

Over the past decade, farm household incomes have surged ahead of average U.S. household incomes (**Figure 11** and **Figure 12**). In 2007 (the last year for which comparable data were available), the average farm household income of $86,233 was nearly 28% higher than the average U.S. household income (**Table 2**).

FARM ASSET VALUES AND DEBT

Farm asset values — which reflect farm investors' and lenders' expectations about long-term profitability of farm sector investments — are projected up nearly 7% in 2008 to a record $2,359 billion, on the strength of continued growth in real estate values. Farm debt is projected nearly flat at $211.7 billion in 2008, up only $0.2 billion from the previous year's record. As a result, farm equity (or net worth, defined as asset value minus debt) is projected at a record $2,147 billion, while the farm debt-to-asset ratio in 2008 is expected to decline to a 49-year low of 9.0%. The U.S. farm debt-to-asset ratio peaked in 1985 at 23%.

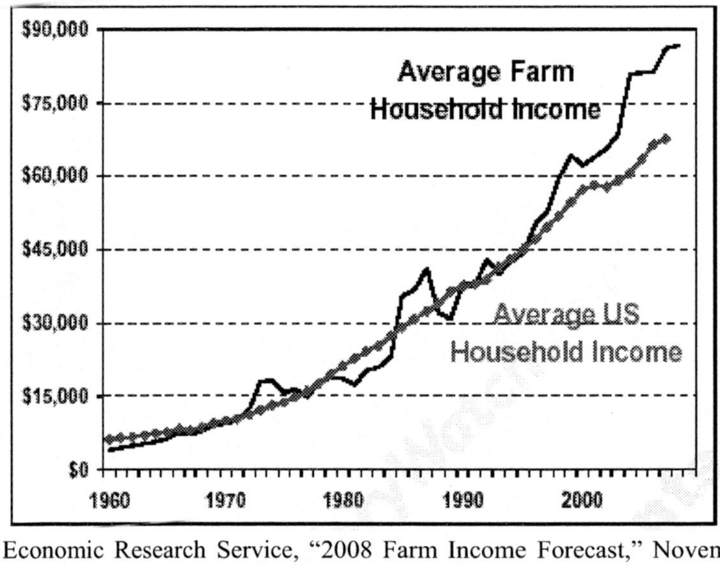

Source: USDA, Economic Research Service, "2008 Farm Income Forecast," November 25, 2008, at http://www.ers.usda.gov/Briefing/FarmIncome/.

Note: 2008 is projected.

Figure 11. Comparison of Farm to U.S. Average Household Income

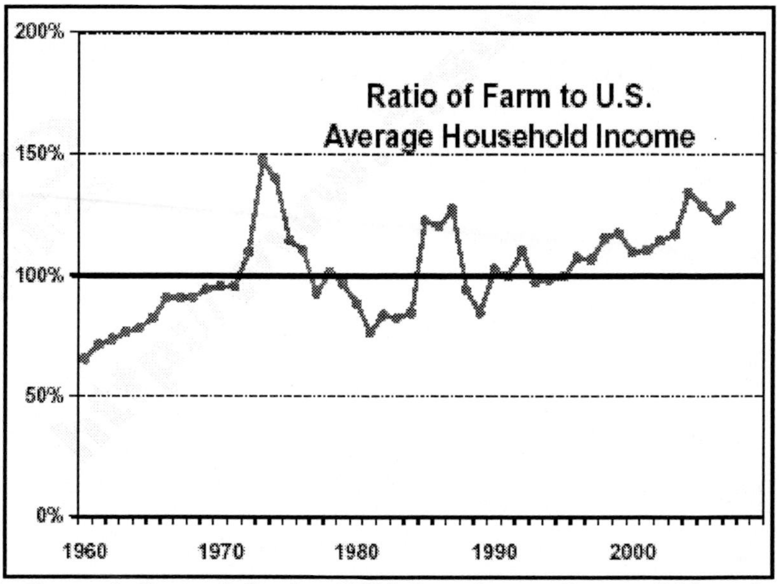

Source: USDA, Economic Research Service, "2008 Farm Income Forecast," November 25, 2008, at http://www.ers.usda.gov/Briefing/FarmIncome/.

Note: 2008 is projected.

Figure 12. Ratio of Farm to U.S. Average Household Income

Table 2. Average Annual Income per U.S. Household, Farm versus All, 2002-2008

	2002	2003	2004	2005	2006	2007	2008
Average U.S. Farm Income by Source				($ per household)			
On-Farm Income	$3,477	$7,884	$13,564	$13,996	$8,750	$8,605	$5,901
Off-Farm income	$62,284	$60,713	$67,279	$67,091	$72,502	$77,618	$80,897
Total Farm income	$65,761	$68,597	$80,843	$81,086	$81,251	$86,223	$86,798
Average U.S. Household Income	$57,852	$59,067	$60,466	$63,344	$66,570	$67,609	na
Farm Household Income as Share of				(percent)			
U.S. Average Household Income	114%	116%	134%	128%	122%	128%	na

Source: USDA, ERS Briefing Room: Farm Household Economics and Well-Being: Historic Data On Farm Operator Household Income, at http://www.ers.usda.gov/Briefing/ WellBeing/Gallery/historic.htm.

Table 3. Average Annual Farm Sector Debt-to-Asset Ratio, 2002-2008F

	2002	2003	2004	2005	2006	2007	2008
($ billions)							
Farm Assets	$1,304.0	1,378.8	1,617.6	1,835.5	2,047.4	2,209.9	2,359.0
Farm Debt	193.3	175.1	183.0	193.2	196.4	211.5	211.7
Farm Equity	1,110.7	1,203.6	1,434.6	1,642.2	1,851.0	1,998.4	2,147.3
(percent)							
Debt-to-Asset Ratio	14.8%	12.7%	11.3%	10.5%	9.6%	9.6%	9.0%

Source: USDA, ERS Briefing Room: Farm Household Economics and Well-Being: Farm Business Balance Sheet,, at http://www.ers.usda.gov/data/FarmBalanceSheet/fbsdmu.htm.

Note: 2008 is projected.

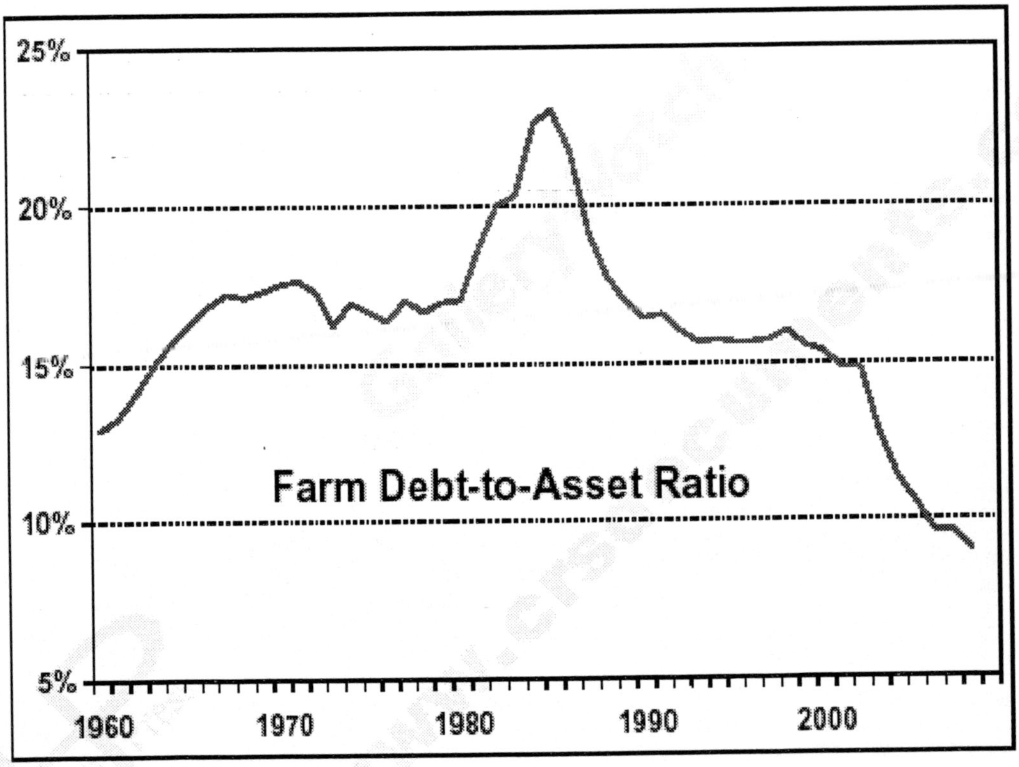

Source: USDA, Economic Research Service, “2008 Farm Income Forecast,” November 25, 2008, at http://www.ers.usda.gov/Briefing/FarmIncome/.

Note: 2008 is projected.

Figure 13. U.S. Farm Debt-to-Asset Ratio Since 1960

Table 4. U.S. Prices and Support Rates for Selected Farm Commodities Since 2002

Commoditya	Unit	Year	2002/03	2003/04	2004/05	2005/06	2006/07	2007/08	2008/09F[b]	% Change[c]	2009/10F	Loan rate[d]	Target Price
Wheat	$/bu	Jun-May	3.56	3.40	3.40	3.42	4.26	6.48	6.50-6.90	3.4%	—	2.75	3.92
Corn	$/bu	Sep-Aug	2.32	2.42	2.06	2.00	3.04	4.20	3.565-4.25	–4.8%	—	1.95	2.63
Sorghum	$/bu	Sep-Aug	2.32	2.39	1.79	1.86	3.29	4.08	2.90-3.50	-19.1%	—	1.95	2.57
Barley	$/bu	Jun-May	2.72	2.83	2.48	2.53	2.85	4.02	4.95-5.35	-46.8%	—	1.85	2.44
Oats	$/bu	Jun-May	1.81	1.48	1.48	1.63	1.87	2.63	2.90-3.10	10.3%	—	1.33	1.44
Rice	$/cwt	Aug-Jul	4.49	8.08	7.33	7.65	9.96	12.80	16.50-17.50	22.3%	—	6.50	10.50
Soybeans	$/bu	Sep-Aug	5.53	7.34	5.74	5.66	6.43	10.10	8.50-9.50	–10.9%	—	5.00	5.80
Soybean oil	¢/lb	Oct-Sep	22.0	30.0	23.0	23.4	31.0	52.0	32.0-35.0	–36.6%	—	—	—
Soybean meal	$/st	Oct-Sep	181 6	256.1	182.9	174.2	205.4	335.9	250-310	–19.6%	—	—	—
Cotton, Upland	¢/lb	Aug-Jul	44.5	61.8	41.6	47.7	46.5	59.3	44.0-52.0	–22.4%	—	52.00	71.25
Choice Steers	$/cwt	Jan-Dec	67.0	84.7	84.8	87.3 85.4	91.8	92.27	0.8%	91-97	—	—	0.8%
Barrows/Gilts	$/cwt	Jan-Dec	34.9	39.5	52.5	50.1	47.3	47.1	47.84	1.3%	47-51	—	—
Broilers	¢/lb	Jan-Dec	55.6	62.0	74.1	70.8	64.4	76.4	79.7	4.3%	81-87	—	—
Eggs	¢/doz	Jan-Dec	67.1	87.9	82.2	65.5	71.8	114.4	127.7	11.6%	118-127	—	—
Milk	$/cwt	Jan-Dec	12.11	12.52	16.05	15.14	12.90	19.13	18.30–18.40	–4.1%	11.80	—	—

Source: Various USDA agency sources as described in the notes below.

a. Season average farm price for grains and oilseeds are from USDA, National Agricultural Statistical Service, Agricultural Prices . Calendar year data is for the first year, e g., 2000/2001 = 2000; F = forecast from World Agricultural Supply and Demand Estimates (WASDE) January 12, 2009; — = no value ; and USDA's out-year 2009/2010 crop price forecasts will first appear in the May 2009 WASDE report. WASDE reports are available at http://www.usda.gov/oce/commodity/wasde/. Soybean and livestock product prices are from USDA, Agricultural Marketing Service (AMS): soybean oil—Decatur, IL, cash price, simple average crude; soybean meal—Decatur. IL, cash price, simple average 48% protein; choice steers—Nebraska, direct 1100-1300 lbs.;

barrows/gilts—national base, live equivalent 51%-52% lean; broilers—wholesale, 12-city average; eggs—Grade A, New York, volume buyers; and milk—simple average of prices received by farmers for all milk.

b. Data for 2008/09 and 2009/10 are USDA forecasts.

c. Percent change from 2007/08, calculated using the difference from the midpoint of the range for 2008/09 with the estimate for 2007/08.

d. Loan rate and target prices are for the 2008/09 crop year. For more information, see CRS Report RL34594, Farm Commodity Programs in the 2008 Farm Bill, by Jim Monke.

End Notes

[1] For information on state-level farm income, see the "U.S. and State Farm Income Data," available as part of the Farm Income Data Files, Farm Income and Costs Briefing Room, Economic Research Service (ERS), USDA, at http://www.ers.usda.gov/data/FarmIncome/finfidmu.htm.

[2] For a more detailed discussion and presentation of statistics on the issues discussed in this report, see Agricultural Income and Finance Outlook, AIS-86, ERS, USDA, December 2008.

[3] ERS's 2008 farm sector income forecast, last updated on August 28, 2008, is available at the Farm Income and Costs Briefing Room, at http://www.ers.usda.gov/Briefing/FarmIncome/nationalestimates.htm.

[4] Production capacity estimates, Renewable Fuels Association, at http://www.ethanolrfa.org/industry/locations/.

[5] For more information see CRS Report RL32712, *Agriculture-Based Renewable Energy Production*, by Randy Schnepf.

[6] USDA, World Agricultural Outlook Board, World Agricultural Supply and Demand Estimates, Jan. 12, 2008.

[7] For more information on commodity programs, see CRS Report RL34594, *Farm Commodity Programs in the 2008 Farm Bill*, by Jim Monke.

[8] For more information, see CRS Report RL31095, *Emergency Funding for Agriculture: A Brief History of Supplemental Appropriations, FY1989-FY2009*, by Ralph M. Chite; and CRS Report RS21212, *Agricultural Disaster Assistance*, by Ralph M. Chite.

[9] For more information on farm typology, see the ERS Briefing Room, *Farm Household Economics and Well-Being: Farm Operator Household Income Forecasts*, at http://www.ers.usda.gov /Briefing/ WellBeing/farmhouseincome.htm.

INDEX

A

B

E

F

G

H

I

J

K

L

M

Q

R

S

T

U

V

W

Y